SpringerBriefs in Mathematics

SpringerBriefs present concise summaries of cutting-edge research and practical applications across a wide spectrum of fields. Featuring compact volumes of 50 to 125 pages, the series covers a range of content from professional to academic. Briefs are characterized by fast, global electronic dissemination, standard publishing contracts, standardized manuscript preparation and formatting guidelines, and expedited production schedules.

Typical topics might include:

A timely report of state-of-the art techniques A bridge between new research results, as published in journal articles, and a contextual literature review A snapshot of a hot or emerging topic An in-depth case study A presentation of core concepts that students must understand in order to make independent contributions.

SpringerBriefs in Mathematics showcases expositions in all areas of mathematics and applied mathematics. Manuscripts presenting new results or a single new result in a classical field, new field, or an emerging topic, applications, or bridges between new results and already published works, are encouraged. The series is intended for mathematicians and applied mathematicians. All works are peer-reviewed to meet the highest standards of scientific literature.

Titles from this series are indexed by Scopus, Web of Science, Mathematical Reviews, and zbMATH.

More information about this series at http://www.springer.com/series/10030

Vijay Gupta • Michael Th. Rassias

Computation
and Approximation

Vijay Gupta
Department of Mathematics
Netaji Subhas University of Technology
New Delhi, India

Michael Th. Rassias
Department of Mathematics
and Engineering Sciences
Hellenic Military Academy
Vari Attikis, Greece

ISSN 2191-8198   ISSN 2191-8201   (electronic)
SpringerBriefs in Mathematics
ISBN 978-3-030-85562-8   ISBN 978-3-030-85563-5   (eBook)
https://doi.org/10.1007/978-3-030-85563-5

This Springer imprint is published by the registered company Springer Nature Switzerland AG
The registered company address is: Gewerbestrasse 11, 6330 Cham, Switzerland

Preface

The concept of approximation concerning positive and linear operators was initiated with the two well-known results due to Weierstrass and Korovkin. During the last century, after the introduction of Bernstein polynomials, two more important operators—namely Baskakov and Szász-Mirakyan operators—were proposed, which are based on negative binomial and Poisson distributions, respectively. These operators satisfy certain partial differential operators, which simplify the determination of the moments and other important properties of approximation. Furthermore, about five decades ago, May (1976) and Ismail-May (1978) presented a method for the construction of a class of exponential type linear positive operators (l.p.o.). Their class of exponential type operators include some basic operators due to Bernstein, Baskakov, Szász-Mirakyan, Gauss-Weierstrass and Post-Widder as special cases. Except for these, some further operators were constructed, but have not been studied yet by researchers due to their unusual behaviour. Very recently, some work has been conducted on these exponential type operators, which is also featured in the present monograph.

For a broad spectrum of sources dealing with approximation and computation in several mathematical disciplines, the interested readers are referred to the references [1–159].

The present book is divided into three chapters:

In the first chapter, we provide a systematic list and properties of exponential type operators available in the literature. We focus mainly on the exponential type operators which are associated with $x(1 + x)^2$, x^3, $2x^{3/2}$ and $2x^2$. It is well-known that by applying diverse methods, convergence estimates may be obtained. To accomplish this precision, adequate and voluminous analysis is involved. In this chapter we also provide some direct estimates and the rate of convergence on such operators. Additionally, for the operators associated with x^3, we provide the convergence estimates in complex setting. In the last part of this section, we indicate some available semi-exponential type operators.

The second chapter deals with some operators of integral type. We first provide a link between the Kantorovich operators with their original operators. Also, we indicate some results for Kantorovich variants of some operators reproducing affine functions. We present some new general classes of operators introduced very recently. Although these operators are not of exponential type, by studying such operators one can investigate many operators simultaneously, rather than studying them individually. Here, we provide different forms of such Durrmeyer type integral operators which preserve constant functions and the affine function. We also discuss the approximation of some new operators as well as their properties in ordinary and simultaneous approximation.

The third chapter deals with the difference between two operators. Here, we provide general estimates for the difference between operators having the same as well as different fundamental functions. We also study general estimates for the difference of operators having higher-order derivatives. In order to exemplify the theoretical results, we provide the quantitative estimates of the differences between certain operators. We also mention some difference estimates in simultaneous approximation.

New Delhi, India   Vijay Gupta

Athens, Greece   Michael Th. Rassias

Contents

Chapter 1
Exponential Type Operators

Bernstein polynomials are known for their wide ranging applications in a broad spectrum of areas, such as statistics and probability theory, numerical analysis, quantum calculus, image processing, p-adic analysis, approximation theory, solution of differential equations, etc. These are basically exponential type operators. Several generalizations of these operators have been considered and their properties have been thoroughly studied.

The main focus of this chapter is on exponential type operators, which were systematically studied first by May in [118], who discussed direct, inverse and saturation results for the linear combination of such operators. The exponential type operators discussed by May [118] include some well-known operators, viz. Bernstein, Baskakov, Szász-Mirakyan, Post-Widder, Gauss-Weierstrass as special cases. The operators were associated with $p(x)$, which are polynomials of degree at most 2.

Later Ismail-May [103] extended this investigation and they raised the following question: "*Are there any more exponential type operators, other than the above known examples, when $p(x) = ax^2 + bx + c$*"? In this way they observed that for each polynomial $p(x)$ under certain assumptions and normalization, one can determine a unique operator L_n. Along with the known operators for given $p(x)$, they constructed some new operators which are associated with x^3, $x(1+x)^2$, $2x^{3/2}$ and $x^2 + 1$. These operators were introduced more than four decades ago in [103], but these were not studied by researchers because of their unusual behaviour. In the last 2 years researchers have been attracted towards these operators and studied their approximation properties. In the present chapter, we deal with such operators of exponential type and indicate their moments and other properties. At the end of this chapter we indicate semi-exponential operators.

1.1 Introduction

An operator of the form

$$(L_n f)(x) = \sum_k \psi_{n,k}^L(x) f\left(\frac{k}{n}\right) \quad \text{or} \quad \int_{-\infty}^{\infty} \phi_n^L(x,t) f(t) dt, \qquad (1.1.1)$$

whose kernels satisfy the partial differential equations

$$\frac{\partial}{\partial x} \psi_{n,k}^L(x) = \frac{k - nx}{p(x)} \psi_{n,k}^L(x) \qquad (1.1.2)$$

$$\frac{\partial}{\partial x} \phi_n^L(x,t) = \frac{n(t-x)}{p(x)} \phi_n^L(x,t), \qquad (1.1.3)$$

respectively, are called exponential type operators. Here $p(x)$ is analytic and positive for $x \in (a,b)$ for some a,b such that $-\infty < a < b < \infty$ and the normalization conditions $(L_n e_0)(x) = e_0$, $e_i(x) = x^i$, $i = 0,1,2,\ldots$ in both forms are satisfied. May [118] and Ismail-May [103] studied the exponential type operators first. The following proposition was proved in [103].

Proposition 1.1 *The partial differential equations given after (1.1.1) along with normalized condition define at most one kernel.*

Some important examples of exponential type operators are the following:
The Gauss-Weierstrass operators (see [29]) for $x \in (-\infty, \infty)$ are defined as:

$$(W_n f)(x) = \sqrt{\frac{n}{2\pi}} \int_{-\infty}^{\infty} \exp\left(\frac{-n(t-x)^2}{2}\right) f(t) dt. \qquad (1.1.4)$$

The Bernstein polynomials (see [29]) for $x \in [0,1]$ are defined as:

$$(B_n f)(x) = \sum_{k=0}^{n} p_{n,k}(x) f\left(\frac{k}{n}\right), \qquad (1.1.5)$$

where $p_{n,k}(x) = \binom{n}{k} x^k (1-x)^{n-k}$ and $p_{n,k}(x) = 0$ if $k < 0$ or $k > n$.
The Baskakov operators (see [26]), for $x \in [0, \infty)$ are defined as

$$(V_n f)(x) = \sum_{k=0}^{\infty} v_{n,k}(x) f\left(\frac{k}{n}\right), \qquad (1.1.6)$$

where $v_{n,k}(x) = \binom{n+k-1}{k} \frac{x^k}{(1+x)^{n+k}}$ is the Baskakov basis function.

The Szász-Mirakyan operators (see [153], [127]), for $x \in [0, \infty)$ are defined as

$$(S_n f)(x) = \sum_{k=0}^{\infty} s_k(nx) f\left(\frac{k}{n}\right), \tag{1.1.7}$$

where $s_k(nx) = e^{-nx} \dfrac{(nx)^k}{k!}$ is the Szász basis function.

The Post-Widder operators (see [157]) for $x \in (0, \infty)$ are defined as

$$(P_n f)(x) = \frac{n^n}{\Gamma(n)x^n} \int_0^{\infty} e^{-nt/x} t^{n-1} f(t) dt. \tag{1.1.8}$$

The Ismail-May operators (see [103, 3.10]) for $x \in (-\infty, \infty)$ are defined as

$$(T_n f)(x) = \frac{n2^{n-2}}{\pi \Gamma(n)} (1+x^2)^{-n/2} \int_0^{\infty} e^{nt \arctan x} \left| \Gamma\left(\frac{n}{2} + \frac{int}{2}\right) \right|^2 f(t) dt. \tag{1.1.9}$$

The Ismail-May operators (see [103, 3.11]) for $x \in (0, \infty)$ are defined as

$$(Q_n f)(x) = \left(\frac{n}{2\pi}\right)^{1/2} e^{n/x} \int_0^{\infty} t^{-3/2} exp\left(-\frac{nt}{2x^2} - \frac{n}{2t}\right) f(t) dt. \tag{1.1.10}$$

The Ismail-May operators (see [103, 3.14]) for $x \in [0, \infty)$ are defined as

$$(R_n f)(x) = \sum_{k=0}^{\infty} r_{n,k}(x) f\left(\frac{k}{n}\right), \tag{1.1.11}$$

where the kernel

$$r_{n,k}(x) = e^{-(n+k)x/(1+x)} \frac{n(n+k)^{k-1}}{k!} \left(\frac{x}{1+x}\right)^k.$$

The Ismail-May operators (see [103, 3.16]) for $x \in (0, \infty)$ are defined as

$$(U_n f)(x) = e^{-n\sqrt{x}} \int_0^{\infty} \left(ne^{-nt/\sqrt{x}} t^{-1/2} I_1\left(2n\sqrt{t}\right) + \delta(t)\right) f(t) dt, \tag{1.1.12}$$

where $\delta(\cdot)$ denotes the Dirac delta function and I_1 represents the modified Bessel function of the first kind given by

$$I_1(z) = \sum_{k=0}^{\infty} \frac{\left(\frac{z}{2}\right)^{1+2k}}{k! \Gamma(k+2)}.$$

The operators introduced in [34, (1.3)] (see also [17, pp 313]) for $x \in (0, \infty)$ are the following sequence of operators:

$$(C_n f)(x) = \frac{1}{(2x)^{n/2}\Gamma(n/2)} \int_0^\infty t^{n/2-1} e^{-t/2x} f\left(\frac{t}{n}\right) dt. \qquad (1.1.13)$$

The following table provides the connections of the above exponential type operators with specific values of $p(x)$

S. No.	Exponential operator	$p(x)$
1	Gauss-Weierstrass operators $(W_n f)(x)$	1
2.	Bernstein polynomials $(B_n f)(x)$	$x(1-x)$
3.	Baskakov operators $(V_n f)(x)$	$x(1+x)$
4.	Szász-Mirakyan operators $(S_n f)(x)$	x
5.	Post-Widder operators $(P_n f)(x)$	x^2
6.	Ismail-May operators $(T_n f)(x)$	$x^2 + 1$
7.	Ismail-May operators $(Q_n f)(x)$	x^3
8.	Ismail-May operators $(R_n f)(x)$	$x(1+x)^2$
9.	Ismail-May operators $(U_n f)(x)$	$2x^{3/2}$
10.	Cismaşiu operators $(C_n f)(x)$	$2x^2$

$$(1.1.14)$$

During the last four decades, although several new operators and generalizations of existing operators have been introduced, to our knowledge no new exponential operators have been introduced by researchers other than the above known operators. By using mathematical software it may be possible to capture some new operators. This could be treated as an open problem for researchers working in the area. Ismail and May [103] established characterization functions by the degree of approximation. For the general exponential type operators L_n, the following moment estimates are true:

Proposition 1.2 (See also [19]) *[103] For the exponential operators (1.1.1), we have*

$$(L_n e_1 f)(x) = x(L_n f)(x) + \frac{p(x)}{n}(L_n f)'(x).$$

In particular, we obtain

$$L_n(e_0, x) = 1$$

$$L_n(e_1, x) = x$$

$$L_n(e_2, x) = x^2 + \frac{p(x)}{n}$$

$$L_n(e_3, x) = x^3 + \frac{3xp(x)}{n} + \frac{p(x)(p(x))'}{n^2}$$

$$L_n(e_4, x) = x^4 + \frac{6x^2 p(x)}{n} + \frac{p(x)[3p(x) + 4x(p(x))']}{n^2}$$

$$+ \frac{p(x)[((p(x))')^2 + p(x)(p(x))'']}{n^3}.$$

After the work of Ismail and May [103] many approximation properties were established for the individual operators due to Gauss-Weierstrass, Bernstein, Baskakov, Szász-Mirakyan and Post-Widder. Altomare and Rasa [19] focused upon the study of a class of second order degenerate elliptic operators on unbounded intervals. It was shown that these operators generate strongly continuous semigroups in suitable weighted spaces of continuous functions. Also, they represented the semigroups as limits of iterates of the exponential type operators.

Sato [148] imposed the condition on $p(x)$ that it is a polynomial of degree ≤ 2 without a double zero and satisfies

$$I =: \frac{n^2}{[p(x)]^2} \int_\alpha^\beta \left[\frac{p'(x)}{n} - t + x \right] (t - x)^3 \phi_n^L(x, t) dt \leq C, \quad (1.1.15)$$

where $\alpha = \min(x, x + p'(x)/n)$ and $\beta = \max(x, x + p'(x)/n)$ and C be an absolute constant independent of n and x. Then the following theorem was discussed in [148]:

Theorem 1.1 *Let $p(x)$ satisfies the condition (1.1.15). Then for $0 < \alpha < 2$ the following statements are equivalent:*

1. $f \in Lip_2\alpha$

2. $|(L_n f)(x) - f(x)| \leq C \left[\frac{p(x)}{n} \right]^{\alpha/2}$,

where $Lip_2\alpha = \{ f \in C[a, b] : \omega_2(f, \delta) \leq O(\delta^\alpha), \delta \to 0_+ \}$.

The method for proving the direct part, i.e. (1.) $\to$ (2.) of this theorem follows using the standard procedure of applying Jackson-type inequality, the Steklov means and appropriate estimates of the moments of the operators. In order to prove the inverse part (2.) $\to$ (1.) Sato [148] used the methods which were introduced by Berens and Lorentz [28]. Sato [148] used this theorem to apply the Gauss-Weierstrass operators, Szász-Mirakjan operators, Bernstein polynomials, Baskakov operators and Ismail-May operators T_n. Rathore and Agrawal [145] obtained inverse and saturation theorems for linear combinations of exponential type operators in simultaneous approximation.

In the present monograph, we indicate recent studies on some of the exponential type operators.

1.2 Operators Associated with $x(1 + x)^2$

The other operators due to Ismail-May [103, 3.14], defined by R_n in (1.1.11) are associated with the Jain operators [107], which for $0 < \alpha < \infty, |\beta| < 1$, are defined as follows:

$$(S_{\alpha,\beta}^n f) := \sum_{k=0}^{\infty} \omega_\beta(k,\alpha) f\left(\frac{k}{n}\right), \qquad (1.2.1)$$

where

$$\omega_\beta(k,\alpha) = \frac{\alpha(\alpha + k\beta)^{k-1}}{k!} e^{-(\alpha+k\beta)}.$$

Jain [107] studied the approximation properties by considering $\alpha = nx$ in the above form (1.2.1). As a special case, if $\beta = 0$, the operators reduce to the well-known Szász-Mirakyan operators. But unlike the operators R_n the operators due to Jain are not exponential type operators. The connection between the two operators can be seen by the following relation:

$$(R_n f)(x) := [(S_{\alpha,\beta}^n f)(x)]_{[\alpha=nx/(1+x), \beta=x/(1+x)]}.$$

Remark 1.1 Also, if $(R_n e_m)(x) = \sum_{k=0}^{\infty} r_{n,k}(x) \left(\frac{k}{n}\right)^m$, $m \in \mathbb{N} \cup \{0\}$ and $e_i(x) = x^1, i = o, 1, 2, \ldots$, then the moments satisfy the following recurrence relation:

$$(R_n e_{m+1})(x) = \frac{x(1+x)^2}{n}[(R_n e_m)(x)]' + x(R_n e_m)(x).$$

Recently Abel and Gupta [2] derived a complete asymptotic expansion for the sequence $(R_n f)(x)$ as n tends to infinity.

Let us denote the falling factorial defined by $z^{\underline{k}} = z(z-1)\cdots(z-k+1)$, for $k \in \mathbb{N}$, and $z^{\underline{0}} = 1$ and the rising factorial by $z^{\overline{k}} = z(z+1)\cdots(z+k-1)$, for $k \in \mathbb{N}$, and $z^{\overline{0}} = 1$. The Stirling numbers of the second kind are denoted by $\sigma(r, \ell)$. Obviously that they can be defined by the relation:

$$z^r = \sum_{\ell=1}^{r} \sigma(r, \ell) z^{\underline{\ell}} \qquad (z \in \mathbb{C}, r = 1, 2, \ldots). \qquad (1.2.2)$$

In particular, we have $\sigma(r, \ell) = 0$, if $\ell > r$. In addition, one defines $\sigma(0, 0) = 1$ and $\sigma(r, 0) = 0$, for $r \in \mathbb{N}$. Based on certain auxiliary results provided in [2], the following complete expansion was established:

Theorem 1.2 ([2]) *Let $q \in \mathbb{N}$ and $x > 0$. For each function $f \in C_\gamma(0, \infty)$, (the class of continuous functions f on $[0, \infty)$ satisfying the growth condition $f(t) =$*

$O\left(e^{\gamma t}\right)$ as $t \to \infty$, for some $\gamma > 0$) having a derivative of order $2q$ at x, the Ismail–May operators $(R_n f)(x)$ possess the complete asymptotic expansion

$$(R_n f)(x) = f(x) + \sum_{k=1}^{q} \frac{1}{n^k} a_k(f, x) + o\left(n^{-q}\right) \qquad (n \to \infty),$$

where the coefficients $a_k(f, x)$ are given by

$$a_k(f, x) = \sum_{s=k+1}^{2k} A_{k,s}(x) f^{(s)}(x) \qquad (k \in \mathbb{N})$$

with

$$A_{k,s}(x) = \frac{1}{s!} \sum_{\ell=0}^{k} \sum_{r=k+1}^{s} (-1)^{s-r} \binom{s}{r} \sigma(r, r - \ell) \binom{r - \ell - 1}{k - \ell}$$

$$\times \frac{x^{s-k}}{(1+x)^{r-k}} c_{r-\ell, k-\ell} \left(\frac{x}{1+x}\right).$$

More explicitly, we have

$$(R_n f)(x)$$

$$= f(x) + \frac{x(1+x)^2}{2n} f^{(2)}(x)$$

$$+ \frac{4x(1+x)^3(1+3x) f^{(3)}(x) + 3x^2(1+x)^4 f^{(4)}(x)}{24n^2}$$

$$+ x(1+x)^4 \frac{\left(30x^2 + 20x + 2\right) f^{(4)}(x) + 4x(1+x)(1+3x) f^{(5)}(x)}{48n^3}$$
$$\frac{+x^2(1+x)^2 f^{(6)}(x)}{}$$

$$+ o\left(n^{-3}\right)$$

as $n \to \infty$. As an immediate consequence, we obtain the following Voronovskaja-type result.

Corollary 1.1 *Let $x > 0$ and $\gamma > 0$. If the function $f \in C_\gamma(0, \infty)$ has a second order derivative at x, then*

$$\lim_{n \to \infty} n\left((R_n f)(x) - f(x)\right) = \frac{x(1+x)^2}{2} f''(x).$$

Also the following asymptotic formula for derivatives can be proved:

Theorem 1.3 *Let $f \in C(a,b)$ such that $|f(x)| \le Me^{N|x|}$ for some $M > 0$. If for some $x \in (a,b)$, f''' exists, then*

$$\lim_{n\to\infty} n((R_n f)'(x) - f'(x)) = \frac{(1+3x)(1+x)}{2} f''(x) + \frac{x(1+x)^2}{2} f'''(x).$$

By changing the variable in the operators R_n the equivalent form for $x \in (0,1)$, was considered in [103] as follows

$$(\widetilde{R}_n f)(x) = \sum_{k=0}^{\infty} \widetilde{r}_{n,k}(x) f\left(\frac{k}{n+k}\right), \qquad (1.2.3)$$

where

$$\widetilde{r}_{n,k}(x) = e^{-nx} \frac{n(n+k)^{k-1}}{k!} (xe^{-x})^k.$$

Unlike the form (1.1.11), the operators $(\widetilde{R}_n)(x)$ as such are not exponential type operators, but can be seen as special kind of exponential type operators (right side coefficient of basis function depending on k), which satisfy

$$x\widetilde{r}'_{n,k}(x) = [k - (n+k)x]\widetilde{r}_{n,k}(x).$$

Using Lagrange's formula, we can write

$$\phi(z) = \phi(0) + \sum_{k=1}^{\infty} \frac{z^k}{(f(z))^k \cdot k!} \left[\frac{d^{k-1}}{du^{k-1}} [(f(u))^k \phi'(u)] \right]_{u=0}$$

and subsequently setting $\phi(z) = e^{nz}$ and $f(z) = e^z$ with $|z| < 1$, to obtain

$$e^{nz} = 1 + \sum_{k=1}^{\infty} n(n+k)^{k-1} \frac{z^k e^{-kz}}{k!} = \sum_{k=0}^{\infty} n(n+k)^{k-1} \frac{z^k e^{-kz}}{k!}. \qquad (1.2.4)$$

Very recently, Lipi and Deo [111] studied the operators $(\widetilde{R}_n f)$. They just indicated the moments, without outlining the proof. Below we discuss the methods to obtain moments. Additionally, one may obtain the moments of higher order using similar reasoning.

Lemma 1.1 *By simple computation using (1.2.4), we have*

$$\sum_{k=0}^{\infty} \widetilde{r}_{n,k}(x) \frac{k(k-1)(k-2)\ldots(k-r)}{(n+k)^{r+1}} = \frac{nx^{r+1}}{n+r+1}.$$

Lemma 1.2 *The few moments of the operators $\widetilde{R}_n$ are given by*

$$(\widetilde{R}_n e_0)(x) = 1$$

$$(\widetilde{R}_n e_1)(x) = \frac{nx}{n+1}$$

$$(\widetilde{R}_n e_2)(x) = \frac{n^2 x^2}{(n+1)(n+2)} + \frac{nx}{(n+1)^2}$$

$$(\widetilde{R}_n e_3)(x) = \frac{n^3 x^3}{(n+1)(n+2)(n+3)} + \frac{n^2 x^2 (3n+4)}{(n+1)^2(n+2)^2} + \frac{nx}{(n+1)^3}$$

$$(\widetilde{R}_n e_4)(x) = \frac{n^4 x^4}{(n+1)(n+2)(n+3)(n+4)} + \frac{2n^3 x^3 (3n^2 + 11n + 9)}{(n+1)^2(n+2)^2(n+3)^2}$$

$$+ \frac{n^2 x^2 (7n^2 + 18n + 12)}{(n+1)^3(n+2)^3} + \frac{nx}{(n+1)^4}.$$

Proof The first two moments are obvious by Lemma 1.1. We start with the second moment as follows

$$(\widetilde{R}_n e_2)(x) = \sum_{k=0}^{\infty} \widetilde{r}_{n,k}(x) \frac{k^2}{(n+k)^2}$$

$$= \sum_{k=0}^{\infty} \widetilde{r}_{n,k}(x) \left[\frac{n}{n+1} \frac{k(k-1)}{(n+k)^2} + \frac{1}{n+1} \frac{k}{(n+k)} \right]$$

$$= \frac{n^2 x^2}{(n+1)(n+2)} + \frac{nx}{(n+1)^2}.$$

Subsequently, we have

$$(\widetilde{R}_n e_3)(x) = \sum_{k=0}^{\infty} \widetilde{r}_{n,k}(x) \frac{k^3}{(n+k)^3}$$

$$= \sum_{k=0}^{\infty} \widetilde{r}_{n,k}(x) \left[\frac{n^2}{(n+1)(n+2)} \frac{k(k-1)(k-2)}{(n+k)^3} \right.$$

$$\left. + \frac{(3n^2 + 4n)}{(n+1)^2(n+2)} \frac{k(k-1)}{(n+k)^2} + \frac{1}{(n+1)^2} \frac{k}{(n+k)} \right]$$

$$= \frac{n^3 x^3}{(n+1)(n+2)(n+3)} + \frac{n^2 x^2 (3n+4)}{(n+1)^2(n+2)^2} + \frac{nx}{(n+1)^3}.$$

We may point out here that the operators can be defined at $x = 0$ but at the point $x = 1$ these operators are not defined, which can also be seen from (1.2.4). In [111, 128] the authors proved some direct results on these operators and their Kantorovich variant for [0, 1]. It is better to consider the interval [0, 1) for these operators.

1.3   Operators Associated with x^3

This section deals with the approximation methods of the operators [103, (3.11)], defined for $x \in (0, \infty)$ as follows:

$$(Q_n f)(x) = \int_0^\infty \phi_n^Q(x, t) f(t) dt, \tag{1.3.1}$$

where the kernel is given by

$$\phi_n^Q(x, t) = \left(\frac{n}{2\pi}\right)^{1/2} e^{n/x} t^{-3/2} exp\left(-\frac{nt}{2x^2} - \frac{n}{2t}\right).$$

These operators satisfy

$$\frac{\partial}{\partial x}\phi_{n,k}^Q(x) = \frac{n(t - x)}{x^3}\phi_{n,k}^Q(x).$$

Some basic lemmas and results have been discussed recently on these operators.

Lemma 1.3 ([66]) *If we denote*

$$T_{n,m}^Q(x) = (Q_n e_m)(x), e_m(t) = t^m, m \in \mathbb{N} \cup \{0\},$$

then the following recurrence relation holds true:

$$n T_{n,m+1}^Q(x) = x^3 [T_{n,m}^Q(x)]' + nx T_{n,m}^Q(x).$$

Also, in the general case, the r-th order moment satisfies the following relation:

$$T_{n,r}^Q(x) = x^r + \frac{r(r - 1)}{2n}x^{r+1} + O(n^{-2}).$$

Lemma 1.4 (See [103, Prop. 2.2], [66]) *If the central moments are defined by* $\mu_{n,m}^Q(x) = (Q_n(t-x)^m)(x), m = 0, 1, 2, \ldots$, *then we have the following recurrence relation*

$$\mu_{n,m+1}^Q(x) = \frac{x^3}{n}\left(\mu_{n,m}^Q(x)\right)' + \frac{mx^3}{n}\mu_{n,m-1}^Q(x).$$

In particular

$$\mu_{n,0}^Q(x) = 1, \quad \mu_{n,1}^Q(x) = 0, \quad \mu_{n,2}^Q(x) = \frac{x^3}{n},$$

$$\mu_{n,3}^Q(x) = \frac{3x^5}{n^2}, \quad \mu_{n,4}^Q(x) = \left(\frac{3x^6}{n^2} + \frac{15x^7}{n^3}\right).$$

Furthermore, for all $x \in (0, \infty)$, we have $\mu_{n,m}^Q(x) = O_x(n^{-[(m+1)/2]})$, where $[\beta]$ stands for the integral part of β.

Let $C_B(0, \infty)$ be the space of all continuous and bounded functions on $(0, \infty)$ endowed with the norm

$$\|f\| = \sup\{|f(x)| : x \in (0, \infty)\}.$$

Theorem 1.4 ([66]) *If $f \in C_B(0, \infty)$, then we have*

$$|(Q_n f)(x) - f(x)| \le C\omega_2\left(f, \frac{x^{3/2}}{\sqrt{n}}\right).$$

Lemma 1.5 ([66]) *It has been observed that, the moment generating function of the operators Q_n is given by*

$$(Q_n e^{At})(x) = \exp\left(\frac{n}{x}\left(1 - \sqrt{\frac{n - 2x^2 A}{n}}\right)\right).$$

Also, we have

$$(Q_n t e^{At})(x) = \frac{n^{1/2}x}{(n - 2Ax^2)^{1/2}} \exp\left(\frac{n}{x}\left(1 - \sqrt{\frac{n - 2x^2 A}{n}}\right)\right)$$

and

$$(Q_n t^2 e^{At})(x) = \left[\frac{x^3}{[n(n - 2Ax^2)]^{1/2}} + x^2\right]$$

$$\times \frac{n}{(n - 2Ax^2)} \exp\left(\frac{n}{x}\left(1 - \sqrt{\frac{n - 2x^2 A}{n}}\right)\right).$$

In the proof of following two theorems, the above Lemma 1.5 will be required as main tool.

Theorem 1.5 ([66]) *The sequence of exponential operators*

$$Q_n : \hat{C}(0, \infty) \to \hat{C}(0, \infty),$$

where $\hat{C}(0, \infty)$ *denotes the class of all real-valued continuous functions* $f(x)$ *having finite value of limit as* $x \to \infty$ *and equipped with uniform norm* $\|.\|_\infty$ *satisfy the following*

$$\|(Q_n f) - f\|_{(0,\infty)} \le 2\omega^* \left(f, \sqrt{2\alpha_1(n) + \alpha_2(n)} \right), f \in \hat{C}(0, \infty),$$

where $\alpha_1(n), \alpha_2(n)$ *tend to zero for n large enough and the modulus of continuity (see [102]), for every* $\delta \ge 0$ *is given by*

$$\omega^*(f, \delta) = \sup_{\substack{x,t \ge 0 \\ |e^{-x} - e^{-t}| \le \delta}} |f(x) - f(t)|.$$

Theorem 1.6 ([66]) *Let* $f, f'' \in \hat{C}(0, \infty)$, *then, for* $x \in (0, \infty)$, *the following inequality holds:*

$$\left| n \left[(Q_n f)(x) - f(x) \right] - \frac{x^3}{2} f''(x) \right|$$

$$\le 2\omega^*(f'', n^{-1/2}) \left[x^3 + \left(\frac{6x}{n} + 3x^6 \right)^{1/2} \left[n^2 \left(Q_n \left(e^{-x} - e^{-t} \right)^4 \right) (x) \right]^{1/2} \right],$$

where $\hat{C}(0, \infty)$ *and the modulus of continuity* $\omega^*(f, \delta)$ *are defined in Theorem 1.5.*

Theorem 1.7 ([66]) *Let* $Q_n : E \to C(0, \infty)$, *where E stands for the space of functions f with exponential growth. If* $f \in C^2(0, \infty) \cap E$ *and* $f'' \in Lip(\beta, A)$, $0 < \beta \le 1$, *then for* $n > 2Ax$ *and* $x \in (0, \infty)$, *we have*

$$\left| (Q_n f)(x) - f(x) - \frac{x^3}{2n} f''(x) \right|$$

$$\le \left[2e^{2Ax} + C(A, x) + \sqrt{C(2A, x)} \right] \cdot \frac{x^3}{2n} \cdot \omega_1 \left(f'', \sqrt{\frac{3x^3}{n} + \frac{15x^4}{n^2}}, A \right),$$

where

$$C(A, x) = \frac{e^{2Ax}}{\left(1 - 2Ax^2 \right)^{3/2}}.$$

The modulus of continuity is given by

$$\omega_1(f, \delta, A) = \sup_{|h| \le \delta, 0 < x < \infty} |f(x) - f(x+h)|e^{-Ax}$$

and the spaces $Lip(\beta, A), 0 < \beta \le 1$ *consist of all those functions satisfying* $\omega_1(f, \delta, A) \le M\delta^\beta$ *for all* $\delta < 1$.

The following lemma is the Lorentz type lemma, which is required for the study of simultaneous approximation.

Lemma 1.6 ([66]) *) There exist polynomials* $q_{i,j,r}(x)$ *independent of* n *and* t *such that*

$$x^{3r} \frac{\partial^r}{\partial x^r}\left[\phi_n^Q(x, t)\right] = \sum_{\substack{2i+j \le r \\ i,j \ge 0}} n^{i+j}(t-x)^j q_{i,j,r}(x)\left[\phi_n^Q(x, t)\right].$$

Using Lemma 1.6, we have the following asymptotic formula.

Theorem 1.8 ([66]) *Let* $f \in C(0, \infty)$ *with* $f(t) = O(e^{\gamma t})$, *for some* $\gamma > 0$, *having the derivative of* $(r+2)$-*th order at a fixed point* $x \in (0, \infty)$. *Then we have*

$$\lim_{n \to \infty} n[(Q_n f)^{(r)}(x) - f^{(r)}(x)] = \frac{r(r-1)(r-2)}{2} f^{(r-1)}(x)$$
$$+ \frac{3r(r-1)x}{2} f^{(r)}(x)$$
$$+ \frac{3rx^2}{2} f^{(r+1)}(x) + \frac{x^3}{2} f^{(r+2)}(x).$$

Additionally, very recently, Gupta [63] estimated the rate of convergence for functions of bounded variation. The main result is based on certain lemmas on moments and the following upper bound:

Lemma 1.7 ([63]) *For each* $x \in (0, \infty)$, *we have*

$$\int_0^x \phi_n^Q(x, t)dt \le \frac{1}{2} + \frac{\sqrt{x}}{2\sqrt{2\pi n}}.$$

Theorem 1.9 ([63]) *Let* f *be a function of bounded variation on each finite subinterval of* $(0, \infty)$ *satisfying the growth condition* $f(t) = O(e^{\gamma t})$ *as* $t \to +\infty$.

Then, for n large, it holds

$$\left| (Q_n f)(x) - \frac{1}{2}(f(x+) + f(x-)) \right|$$

$$\leq \frac{\sqrt{x}}{2\sqrt{2\pi n}} |f(x+) - f(x-)| + \frac{2x+1}{n} \sum_{k=1}^{n} V_{x-x/\sqrt{k}}^{x+x/\sqrt{k}}(f_x)$$

$$+ \frac{x e^{\gamma x}}{n} + \frac{\sqrt{x}}{\sqrt{n}} \left(\exp \left(2\gamma x + \frac{x^3(2\gamma)^2}{2n} + \frac{x^5(2\gamma)^3}{2n^2} + \cdots \right) \right)^{1/2},$$

where

$$f_x(t) = \begin{cases} f(t) - f(x-), & 0 < t < x \\ f(t) - f(x+), & x < t < \infty \\ 0, & t = x. \end{cases}$$

In recent years, the study of approximation by complex operators on compact disks constitutes an active area of research. The commendable work has been done in complex setting by Gal et al., cf. [52, 53, 55]. In earlier work such operators were not discussed. Gal and Gupta [54] in their recent paper have established some results in complex setting on the operators $Q_n(f, z)$. Their main results are based on moment estimation and the following lemmas:

Lemma 1.8 *Suppose that* $f : \mathbb{C} \to \mathbb{C}$, $f(z) = \sum_{k=0}^{\infty} c_k z^k$, *is an entire function satisfying the condition* $|c_k| \leq M \frac{A^k}{k!}$, $k = 0, 1, \ldots$, *with* $M > 0$ *and* $A \in (0, 1/2)$ *(which implies that* f *is of exponential growth since* $|f(z)| \leq M \cdot \exp(A|z|)$, *for all* $z \in \mathbb{C}$). *Then,* $Q_n(f, z)$ *is well defined for any* $n \in \mathbb{N}$ *and any* $z \in \mathbb{C}$ *satisfying*

$$\Re(z^2) > 0 \text{ and } \frac{|z|^2}{\Re(z^2)} < \frac{1}{2A}. \tag{1.3.2}$$

Lemma 1.9 *Suppose that* f *is an entire function, i.e.* $f(z) = \sum_{k=0}^{\infty} c_k z^k$ *for all* $z \in \mathbb{C}$ *such that there exist* $M > 0$ *and* $A \in (0, 1)$, *with the property* $|c_k| \leq M \frac{A^k}{k!}$, *for all* $k = 0, 1, \ldots$, *(which implies* $|f(z)| \leq M \exp(A|z|)$ *for all* $z \in \mathbb{C}$).

Then for all $n \in \mathbb{N}$ *and* z *satisfying (1.3.2), we have*

$$Q_n(f, z) = \sum_{k=0}^{\infty} c_k Q_n(e_k, z).$$

Based on the above lemmas, the following upper estimate was provided in [54].

Theorem 1.10 *Suppose that* f *is an entire function, i.e.* $f(z) = \sum_{k=0}^{\infty} c_k z^k$ *for all* $z \in \mathbb{C}$ *such that there exist* $M > 0$ *and* $A \in (0, 1/2)$, *with the property* $|c_k| \leq M \frac{A^k}{k!}$, for all $k = 0, 1, \ldots$, (which implies $|f(z)| \leq Me^{A|z|}$ for all $z \in \mathbb{C}$). Consider $1 \leq r < \frac{1}{A}$.

Then for all $n \geq r^2$, $|z| \leq r$ and z satisfying (1.3.2), the following estimate holds:

$$|Q_n(f, z) - f(z)| \leq \frac{C_{r,M,A}}{n},$$

where

$$C_{r,M,A} = Mr \sum_{k=2}^{\infty} (k + 1)(Ar)^k < \infty.$$

The following Voronovskaja-kind quantitative result was established in [54]:

Theorem 1.11 *Suppose that f is an entire function, i.e. $f(z) = \sum_{k=0}^{\infty} c_k z^k$ for all $z \in \mathbb{C}$ such that there exist $M > 0$ and $A \in (0, 1/2)$, with the property $|c_k| \leq M \frac{A^k}{k!}$, for all $k = 0, 1, \ldots$, (which implies $|f(z)| \leq M \exp(A|z|)$ for all $z \in \mathbb{C}$). Consider $1 \leq r < \frac{1}{A}$.*

Then for all $n \geq r^2$, $|z| \leq r$ and z satisfying (1.3.2), the following estimate hold:

$$\left| Q_n(f, z) - f(z) - \frac{z^3 f''(z)}{2n} \right| \leq \frac{E_{r,M,A}(f)}{n^2}, \text{ where}$$

$$E_{r,M,A}(f) = 3Mr^2 \cdot \sum_{k=2}^{\infty} (k + 1)^2 (Ar)^k < +\infty.$$

Also, by applying the above Voronovskaja-kind theorem, the following lower order in approximation was given in [54]:

Theorem 1.12 *Under the hypothesis in Theorem 1.11, if f is not a polynomial of degree ≤ 1, then for all $n \geq r^2$ we have*

$$\|Q_n(f, \cdot) - f\|_r^* \geq \frac{K_{r,M,A}(f)}{n},$$

where $\|F\|_r^* = \sup\{|F(z)|; |z| \leq r \text{ and } z \text{ satisfies } (1.3.2)\}$ and $K_{r,M,A}(f)$ is a constant which depends only on f, M, A and r.

Combining Theorem 1.10 with Theorem 1.12, we immediately get the following exact estimate.

Corollary 1.2 *Under the hypothesis in Theorem 1.11, if f is not a polynomial of degree ≤ 1, then we have*

$$\|Q_n(f, \cdot) - f\|_r^* \sim \frac{1}{n}, n \in \mathbb{N},$$

where the symbol $\sim$ represents the well-known equivalence between the orders of approximation.

Moreover, for some further recent results on Ismail-May type operators, we refer the interested readers to [112, 129].

1.4   Operators Associated with $2x^{3/2}$

After the work on exponential type operators by May [118], Ismail-May [103, (3.16)] proposed one more operator defined as

$$(U_n f)(x) = \int_0^\infty \phi_n^U(x, t) f(t) dt, \qquad (1.4.1)$$

where

$$\phi_n^U(x, t) = e^{-n\sqrt{x}} \left\{ e^{-nt/\sqrt{x}} \sum_{k=0}^\infty \frac{t^k n^{2(1+k)}}{k!(k+1)!} + \delta(t) \right\},$$

and $\delta(\cdot)$ denotes the Dirac delta function. These operators are connected with $2x^{3/2}$, i.e.

$$2x^{3/2}(U_n f)'(x) = (U_n(e_1 - xe_0)nf)(x).$$

It was observed by Abel and Gupta [1] that the operators U_n are closely related to the well-known Phillips operators given by

$$\left(\widehat{P}_n f\right)(x) = n \sum_{k=0}^\infty s_k(nx) \int_0^\infty s_{k-1}(nt) f(t) dt + e^{-nx} f(0),$$

where $s_k(nx) = e^{-nx} \frac{(nx)^k}{k!}$. The relation between them is the following:

$$\left(\widehat{P}_{n/\sqrt{x}} f\right)(x) = (U_n f)(x),$$

but unlike the operators U_n, the Phillips operators are not exponential type operators. Gupta [67, Remark 1] derived the following formula.

Lemma 1.10 *The moment generating function of the operators U_n is given by*

$$\left(U_n e^{\theta t}\right)(x) = \exp\left(\frac{n\theta x}{n - \theta\sqrt{x}}\right) \qquad (n > |\theta|\sqrt{x}).$$

Expanding in powers of θ, we obtain, for $|\theta| < n/\sqrt{x}$, the following

$$
\left(U_n e^{\theta t}\right)(x) = 1 + x\theta + \left(\frac{2x^{3/2}}{n} + x^2\right)\frac{\theta^2}{2!}
$$

$$
+ \frac{\left(6x^2 + 6nx^{5/2} + n^2 x^3\right)\theta^3}{n^2 \quad 3!}
$$

$$
+ \frac{\left(24x^{5/2} + 36nx^3 + 12n^2 x^{7/2} + n^3 x^4\right)\theta^4}{n^3 \quad 4!} + \cdots
$$

Evidently, the coefficients of $\theta^r / r!$ provide the r-th order moments $(U_n e_r)(x)$. Another method for obtaining moments in terms of the falling factorial has recently been given by Abel et al (see [4]) in the following two lemmas.

Lemma 1.11 *For $r \in \mathbb{N}$, the r-th order moment of the Ismail-May operators U_n has the representation*

$$
(U_n e_r)(x) = \sum_{k=0}^{r-1} \frac{1}{n^k}\binom{r-1}{k} r^{\underline{k}} x^{r-\frac{k}{2}},
$$

where the falling factorial is given by

$$
z^{\underline{m}} = z(z-1)(z-2)\cdots(z-m+1), \text{ for } m \in \mathbb{N} \text{ and } z^{\underline{0}} = 1.
$$

Lemma 1.12 *For $s = 1, 2, \ldots$, the central moment $\left(U_n \psi_x^s\right)(x)$, of the Ismail-May operator U_n possesses the representation*

$$
\left(U_n \psi_x^s\right)(x) = \sum_{k=\lfloor (s+1)/2 \rfloor}^{s-1} n^{-k}\binom{s}{k} k^{\underline{s-k}}(k-1)^{\underline{2k-s}} x^{s-\frac{k}{2}}.
$$

Following basic theorems, these operators have been recently estimated by Gupta [67].

Theorem 1.13 ([67]) *Let $f \in C_B(0, \infty)$, then we have*

$$
|(U_n f)(x) - f(x)| \leq C\omega_2\left(f, \frac{x^{3/4}}{\sqrt{n}}\right).
$$

Theorem 1.14 ([67]) *Let f be bounded and integrable function on the interval $(0, \infty)$, possessing a second derivative of f at a fixed point $x \in (0, \infty)$, then*

$$
\lim_{n \to \infty} n\left((U_n f)(x) - f(x)\right) = x^{3/2} f''(x).
$$

Theorem 1.15 ([67]) *Let* $f, f'' \in \hat{C}(0, \infty)$, *then, for* $x \in (0, \infty)$, *the following inequality holds:*

$$\left| n\left[(U_n f)(x) - f(x) \right] - x^{3/2} f''(x) \right|$$

$$\leq 2\omega^*(f'', n^{-1/2}) \left[2x^{3/2} + \left(\frac{24x^{5/2}}{n} + 12x^3 \right)^{1/2} \left[n^2 (U_n(e^{-x} - e^{-t})^4)(x) \right]^{1/2} \right],$$

where the class $\hat{C}$ *and modulus of continuity are defined in Theorem 1.5.*

In the recent paper [87], Gupta et al. provided the direct estimate based on the following remark:

Remark 1.2 By simple computation, we have

$$(U_n e^{Bt} e_0)(x) = e^{\frac{nBx}{(n-B\sqrt{x})}}$$

$$(U_n e^{Bt} e_1)(x) = e^{\frac{nBx}{(n-B\sqrt{x})}} \frac{n^2 x}{(n - B\sqrt{x})^2}$$

$$(U_n e^{Bt} e_2)(x) = e^{\frac{nBx}{(n-B\sqrt{x})}} \frac{n^2 x^{3/2} \left(2n - 2B\sqrt{x} + n^2 x^{1/2}\right)}{\left(n - B\sqrt{x}\right)^4}.$$

Thus, we have

$$(U_n e^{Bt} (e_2 - e_0 x)^2)(x) \leq C(B, x) \mu_{n,2}(x),$$

where

$$C(B, x) = e^{2Bx}(8 + 17Bx).$$

Theorem 1.16 ([87]) *For the operators* $U_n : E \rightarrow C(0, \infty)$, *where* E *is the space of functions* f *having exponential growth, let*

$$f \in C^2(0, \infty) \cap E \qquad and \qquad f'' \in \text{Lip}(\beta, B) \quad (0 < \beta \leq 1).$$

Then, for $x \in (0, \infty)$ *and* $n > 2B\sqrt{x}$, *it is asserted that*

$$\left| (U_n f)(x) - f(x) - \frac{x^{3/2}}{n} f''(x) \right|$$

$$\leqq \left(e^{2Bx} + \frac{C(B, x)}{2} + \frac{\sqrt{C(2B, x)}}{2} \right) \cdot \frac{2x^{3/2}}{n}$$

$$\cdot \omega_1 \left(f'', \sqrt{\left(\frac{12x}{n^2} + \frac{6x^{3/2}}{n} \right)}, B \right),$$

where the modulus of continuity $\omega_1(f, \delta, B)$ is as considered in Theorem 1.7, $C(B, x)$ is given in Remark 1.2 and the spaces $\mathrm{Lip}(\beta, B)$ $(0 < \beta \leq 1)$ consist of all functions f such that

$$\omega_1(f, \delta, A) \leqq M\delta^\beta \qquad (\forall \, \delta < 1).$$

Along with other direct results, the following asymptotic formula in simultaneous approximation was provided:

Theorem 1.17 ([87]) *Let $f \in C_\gamma(0, \infty)$, the class of continuous functions on $(0, \infty)$ with $|f(t)| \leq Me^{t\gamma}$ $(M > 0)$. Also let $f^{(r+2)}(x)$ exist at a fixed point $x \in (0, \infty)$. Then, for $r = 0, 1, 2$, it is asserted that*

$$\lim_{n \to \infty} n\left[(U_n^{(r)} f)(x) - f^{(r)}(x)\right]$$

$$= x^{-3/2} \frac{f^{(r-1)}(x)}{(r-3)!} \left\{\left(r - \frac{3}{2}\right)\left(r - \frac{5}{2}\right) \cdots \left(-\frac{1}{2}\right)\right\}$$

$$+ x^{-1/2} \frac{f^{(r)}(x)}{(r-2)!} \left[-(r-2)\left\{\left(r - \frac{3}{2}\right)\left(r - \frac{5}{2}\right) \cdots \left(-\frac{1}{2}\right)\right\}\right.$$

$$\left. + \left\{\left(r - \frac{1}{2}\right)\left(r - \frac{3}{2}\right) \cdots \left(\frac{1}{2}\right)\right\}\right]$$

$$+ x^{1/2} \frac{f^{(r+1)}(x)}{(r-1)!} \left[\frac{(r-1)(r-2)}{2} \left\{\left(r - \frac{3}{2}\right)\left(r - \frac{5}{2}\right) \cdots \left(-\frac{1}{2}\right)\right\}\right.$$

$$- (r-1)\left\{\left(r - \frac{1}{2}\right)\left(r - \frac{3}{2}\right) \cdots \left(\frac{1}{2}\right)\right\}$$

$$\left. + \left\{\left(r + \frac{1}{2}\right)\left(r - \frac{1}{2}\right) \cdots \frac{3}{2}\right\}\right]$$

$$+ x^{3/2} \frac{f^{(r+2)}(x)}{r!} \left[-\frac{r(r-1)(r-2)}{6} \left\{\left(r - \frac{3}{2}\right)\left(r - \frac{5}{2}\right) \cdots \left(-\frac{1}{2}\right)\right\}\right.$$

$$+ \frac{r(r-1)}{2} \left\{\left(r - \frac{1}{2}\right)\left(r - \frac{3}{2}\right) \cdots \left(\frac{1}{2}\right)\right\}$$

$$\left. - r\left\{\left(r + \frac{1}{2}\right)\left(r - \frac{1}{2}\right) \cdots \frac{3}{2}\right\} + \left\{\left(r + \frac{3}{2}\right)\left(r + \frac{1}{2}\right) \cdots \frac{5}{2}\right\}\right],$$

where the terms within the curly brackets end with the last terms as indicated in each bracket and, otherwise, its value is 1.

Let $C_\beta(\mathbb{R}^+)$ be the class of continuous functions f on $[0, \infty)$ satisfying the exponential growth condition $f(t) = O\left(e^{\beta t}\right)$ as $t \to \infty$, for some $\beta > 0$.

The complete asymptotic expansion for U_n, derived in [4] is as follows:

Theorem 1.18 ([4]) *Let $q \in \mathbb{N}$ and $x \in (0, \infty)$. For each function $f \in C_\beta (\mathbb{R}^+)$, which has a derivative of order $2q$ at the point x, the operators T_n possess the asymptotic expansion*

$$(U_n f)(x) = f(x) + \sum_{k=1}^{q} \frac{1}{k! n^k} \sum_{s=1}^{k} \binom{k}{s} (k-1)^{\underline{k-s}} f^{(k+s)}(x) x^{s + \frac{k}{2}} + o(n^{-q})$$

as $n \to \infty$.

Additionally, in [4] the operators U_n were appropriately modified to preserve the exponential function e^{Ax} and the modified operators take the form:

$$(\widetilde{U}_n f)(x) = \int_0^\infty \Phi_n^U (a_n(x), t) f(t) dt, \qquad (1.4.2)$$

where

$$a_n(x) = \frac{x(2n^2 + A^2 x) + Ax\sqrt{A^2 x^2 + 4xn^2}}{2n^2}$$

and $\Phi_n^U (a_n(x), t)$ is defined in (1.4.1). For such modification, the asymptotic formula takes the following forms:

Theorem 1.19 ([4]) *Let $f \in C_\beta (\mathbb{R}^+)$ for some $\beta > 0$. If f'' exists at a point $x \in (0, \infty)$, then we have*

$$\lim_{n \to \infty} n \left((\widetilde{U}_n f)(x) - f(x) \right) = Ax^{3/2} f'(x) + x^{3/2} f''(x).$$

Theorem 1.20 ([4]) *Let $f \in C_\beta (\mathbb{R}^+)$ admitting the derivative of 3rd order at a fixed point $x \in (0, \infty)$. We have*

$$\lim_{n \to \infty} n \left[\left(\widetilde{U}_n^{(1)} f \right)(x) - f^{(1)}(x) \right] = \frac{3Ax^{1/2}}{2} f^{(1)}(x) + \left(\frac{3}{2} x^{1/2} + Ax^{3/2} \right) f^{(2)}(x)$$
$$+ x^{3/2} f^{(3)}(x).$$

In continuation, for these operators U_n, Abel and Gupta [1] established the rate of convergence for functions of bounded variation. To estimate the rate of convergence, the following important lemma is required.

Lemma 1.13 *For each $x \in (0, \infty)$, we have*

$$\int_0^x \phi_n^U (x, t) dt = \frac{1}{2} + \frac{1}{4 \sqrt[4]{x} \sqrt{\pi n}} + \frac{1}{64 \sqrt[4]{x^3} \sqrt{\pi} n^{3/2}} + O\left(n^{-2} \right) \qquad (n \to \infty).$$

Using Lemma 1.13 along with other basic results the following main result was estimated.

Theorem 1.21 ([1]) *Let f be a function of bounded variation on each finite subinterval of $(0, \infty)$. Suppose that f satisfies the growth condition $|f(t)| \leq Ce^{\gamma t}$, for $t > 0$. Then, there exists a sequence $(\varepsilon_n(x))$ with $\varepsilon_n(x) \to 0$ as $n \to \infty$, such that, for $n > 2\gamma \sqrt{x}$,*

$$\left| (U_n f)(x) - \frac{1}{2}(f(x+) - f(x-)) \right|$$

$$\leq \left(\frac{1}{2\sqrt[4]{x}\sqrt{\pi n}} + \frac{1}{32\sqrt[4]{x^3}\sqrt{\pi}n^{3/2}} \right) |f(x+) - f(x-)|$$

$$+ \frac{2 + \sqrt{x}}{n\sqrt{x}} \sum_{k=1}^{n} V_{x-x/\sqrt{k}}^{x+x/\sqrt{k}}(f_x) + \frac{2Ce^{\gamma x}}{n\sqrt{x}}$$

$$+ \frac{\sqrt{2}Cx^{3/4}}{\sqrt{n}} \exp\left(\frac{n\gamma x}{n - 2\gamma \sqrt{x}} \right) + \frac{\varepsilon_n(x)}{n^2},$$

where

$$f_x(t) = \begin{cases} f(t) - f(x-) & (0 < t < x), \\ 0 & (t = x), \\ f(t) - f(x+) & (x < t < \infty). \end{cases}$$

1.5   Operators Associated with $2x^2$

One of the operators introduced in [35, (1.3)] (see also [34]) for $x \in (0, \infty)$ is the following sequence of operators:

$$(C_n f)(x) = \frac{1}{(2x)^{n/2}\Gamma(n/2)} \int_0^\infty t^{n/2-1} e^{-t/2x} f\left(\frac{t}{n} \right) dt.$$

Altomare and Campiti in their important book [17, pp 313] termed these operators as Cismaşiu operators. These operators preserve constant and linear functions. Alternatively, we can write C_n as

$$(C_n f)(x) = \frac{n^{n/2}}{(2x)^{n/2}\Gamma(n/2)} \int_0^\infty t^{n/2-1} e^{-nt/2x} f(t) dt. \tag{1.5.1}$$

where the kernel is given by

$$\phi_n^C(x, t) = \frac{n^{n/2}}{(2x)^{n/2}\Gamma(n/2)} t^{n/2-1} e^{-nt/2x}.$$

Altomare and Diomede in [18] estimated some interesting results for weighted approximation and established asymptotic formulae. We observe here that these operators are very closely related to the well-known Post-Widder operators. Actually this operator is nothing else than the well-known gamma operator, which can be obtained by replacing n in (1.5.1) with $2n$. Here we estimate some approximation properties of these operators having exponential growth $f(t) = a^{Bt}, a > 1$.

Lemma 1.14 *It has been observed that The moment generating function of the operators C_n is given by*

$$(C_n a^{Bt})(x) = \left(1 - \frac{2xB\ln a}{n}\right)^{-n/2}.$$

We also have

$$(C_n t a^{Bt})(x) = x\left(1 - \frac{2xB\ln a}{n}\right)^{-\frac{n}{2}-1}$$

and

$$(C_n t^2 a^{Bt})(x) = x^2\left(\frac{n+2}{n}\right)\left(1 - \frac{2xB\ln a}{n}\right)^{-\frac{n}{2}-2}.$$

The proof of the above lemma is simple and we thus omit the details.

Lemma 1.15 *If we denote $T_{n,m}^C(x) = (C_n e_m)(x)$, $e_m(t) = t^m$, $m \in N \cup \{0\}$, then we have*

$$(C_n e^{At})(x) = \left(1 - \frac{2xA}{n}\right)^{-n/2}.$$

The moments are given as

$$T_{n,m}(x) = \left[\frac{\partial^m}{\partial A^m}\left\{\left(1 - \frac{2Ax}{n}\right)^{-n/2}\right\}\right]_{A=0}.$$

Some of the moments are listed below:

$$T_{n,1}^C(x) = x, \ T_{n,2}^C(x) = x^2 + \frac{2x^2}{n}$$

$$T_{n,3}^C(x) = x^3 + \frac{6x^3}{n} + \frac{8x^3}{n^2}, \ T_{n,4}^C(x) = x^4 + \frac{12x^4}{n} + \frac{92x^4}{n^2}.$$

We also observe that the m-th order moment satisfies the following representation:

$$T_{n,m}^C(x) = x^m + \frac{m(m-1)}{n}x^m + O(n^{-2}).$$

Lemma 1.16 *If the central moments are defined by*

$$\mu_{n,m}^C(x) = (C_n(e_1 - xe_0)^m)(x), \ m = 0, 1, 2, \ldots,$$

then

$$\mu_{n,m}^C(x) = \left[\frac{\partial^m}{\partial A^m} \left\{ e^{-Ax} \left(1 - \frac{2Ax}{n} \right)^{-n/2} \right\} \right]_{A=0}.$$

In particular

$$\mu_{n,0}^C(x) = 1, \ \ \mu_{n,1}^C(x) = 0, \ \ \mu_{n,2}^C(x) = \frac{2x^2}{n},$$

$$\mu_{n,3}^C(x) = \frac{8x^3}{n^2}, \ \ \mu_{n,4}^C(x) = \left(\frac{12x^4}{n^2} + \frac{48x^4}{n^3} \right).$$

Furthermore, for all $x \in (0, \infty)$, we have $\mu_{n,m}^C(x) = O_x(n^{-[(m+1)/2]})$, where $[b]$ stands for the integral part of b.

Corollary 1.3 *Let β and δ be positive real numbers with $[a, b] \subset (0, \infty)$. Then for any $s > 0$, we have*

$$\sup_{x \in [a,b]} \left| \int_{|t-x| \geq \delta} \phi_n^C(x, t) a^{\beta t} dt \right| = O(n^{-s}).$$

Theorem 1.22 *If $f \in C_B(0, \infty)$, the space of all continuous and bounded functions on $(0, \infty)$, then we have*

$$|(C_n f)(x) - f(x)| \leq C\omega_2 \left(f, \frac{x}{\sqrt{n}} \right).$$

In the next two main theorems, we use the following modulus of continuity (that follows and generalizes the definition due to [102]), which for every $\delta \geq 0$ is given by

$$\widetilde{\omega}(f, \delta) = \sup_{\substack{|a^{-x}-a^{-t}|\leq\delta \\ x,t\geq 0}} |f(x) - f(t)|.$$

This modulus of continuity has the property

$$|f(t) - f(x)| \leq \left(1 + \frac{(a^{-x} - a^{-t})^2}{\delta^2}\right) \widetilde{\omega}(f, \delta), \quad \delta > 0.$$

Let $L_n : \hat{C}(0, \infty) \rightarrow \hat{C}(0, \infty)$, where $\hat{C}(0, \infty)$ denotes the class of all real-valued continuous functions $f(x)$ having finite value of limit as $x \rightarrow \infty$ and equipped with uniform norm $||.||_\infty$

Theorem 1.23 *The sequence of exponential operators* $C_n : \hat{C}(0, \infty) \rightarrow \hat{C}(0, \infty)$ *satisfy the following*

$$||(C_n f) - f||_\infty \leq 2\widetilde{\omega}\left(f, \sqrt{2\alpha_1(n) + \alpha_2(n)}\right), f \in \hat{C}(0, \infty),$$

where $\alpha_1(n), \alpha_2(n)$ *tend to zero for n large enough.*

Proof The operators preserve the constant function, therefore we have $\alpha_0(n) = 0$. Furthermore

$$(C_n a^{-t})(x) = \left(1 + \frac{2x \ln a}{n}\right)^{-n/2}.$$

Let

$$f_n(x) = \left(1 + \frac{2x \ln a}{n}\right)^{-n/2} - a^{-x}.$$

Since $f_n(0) = f_n(\infty) = 0$, there exists a point $\eta_n \in (0, \infty)$ such that

$$||f_n||_\infty = f_n(\eta_n).$$

Thus $f_n'(\eta_n) = 0$ implies

$$\left(1 + \frac{2\eta_n \ln a}{n}\right)^{-n/2-1} = a^{-\eta_n}.$$

Hence

$$f_n(\eta_n) = \left(1 + \frac{2\eta_n \ln a}{n}\right)^{-n/2} - \left(1 + \frac{2\eta_n \ln a}{n}\right)^{-n/2-1}$$

$$= \left(1 + \frac{2\eta_n \ln a}{n}\right)^{-n/2-1}\left[\left(1 + \frac{2\eta_n \ln a}{n}\right) - 1\right] \to 0$$

as $n \to \infty$.

Finally

$$(C_n a^{-2t})(x) = \left(1 + \frac{4x \ln a}{n}\right)^{-n/2}.$$

Consider

$$g_n(x) = \left(1 + \frac{4x \ln a}{n}\right)^{-n/2} - a^{-2x}.$$

Since $g_n(0) = g_n(\infty) = 0$, there exists a point $\zeta_n \in (0, \infty)$ such that

$$\|g_n\|_\infty = g_n(\zeta_n).$$

Thus $g_n'(\zeta_n) = 0$ implies

$$\left(1 + \frac{4\zeta_n \ln a}{n}\right)^{-n/2-1} = a^{-2\zeta_n}.$$

Therefore

$$g_n(\zeta_n) = \left(1 + \frac{4\zeta_n \ln a}{n}\right)^{-n/2} - \left(1 + \frac{4\zeta_n \ln a}{n}\right)^{-n/2-1}$$

$$= \left(1 + \frac{4\zeta_n \ln a}{n}\right)^{-n/2-1}\left[\left(1 + \frac{4\zeta_n \ln a}{n}\right) - 1\right] \to 0$$

as $n \to \infty$.   ∎

Theorem 1.24 *Let $f, f'' \in \hat{C}(0, \infty)$, then, for $x \in (0, \infty)$, the following inequality holds:*

$$\left| n\left[(C_n f)(x) - f(x)\right] - x^2 f''(x) \right|$$

$$\leq 4\tilde{\omega}(f'', n^{-1/2})\left[x^2 + \left(\frac{16x^4}{n} + 3x^4\right)^{1/2}\left[A_{n,x,a}\right]^{1/2}\right],$$

where $A_{n,x,a} = n^2(C_n \left(a^{-x} - a^{-t}\right)^4)(x)$.

Proof By Taylor's expansion, we have

$$f(t) = \sum_{i=0}^{2} (t-x)^i \frac{f^{(i)}(x)}{i!} + \frac{f''(\xi) - f''(x)}{2} (t-x)^2 ,$$

where ξ lying between x and t. Applying the operator C_n to above equality, we can write that

$$\left| (C_n f)(x) - \mu_{n,0}^C(x) f(x) - \mu_{n,1}^C(x) f'(x) - \tfrac{1}{2} \mu_{n,2}^C(x) f''(x) \right|$$
$$= \left| \left(C_n \frac{f''(\xi)-f''(x)}{2} (t-x)^2 \right)(x) \right|.$$

Thus, using Lemma 1.16, we get

$$\left| n\left[(C_n f)(x) - f(x) \right] - x^2 f''(x) \right| = \left| n \left(C_n \frac{f''(\xi)-f''(x)}{2} (t-x)^2 \right)(x) \right|.$$

Using similar methods to those in [38, Th. 2], we can write

$$\left| \frac{f''(\xi) - f''(x)}{2} \right| \leq 2 \left(1 + \frac{(a^{-x} - a^{-t})^2}{\delta^2} \right) \widetilde{\omega}(f'', \delta).$$

Hence, after applying the Cauchy-Schwarz inequality we get

$$n\, C_n \left(\left| \frac{f''(\xi) - f''(x)}{2} \right| (t-x)^2, x \right)$$
$$\leq 2n\, \widetilde{\omega}(f'', \delta) \left[\mu_{n,2}^C(x) + \frac{1}{\delta^2} \sqrt{(C_n(a^{-x} - a^{-t})^4)(x)} \sqrt{\mu_{n,4}^C(x)} \right].$$

Considering $\delta = n^{-1/2}$, we obtain

$$n \left(C_n \left| \frac{f''(\xi) - f''(x)}{2} \right| (t-x)^2 \right)(x)$$
$$\leq 2\widetilde{\omega}\left(f'', \frac{1}{\sqrt{n}} \right) \left[n\mu_{n,2}^C(x) + \sqrt{A_{n,x,a}} \sqrt{n^2 \mu_{n,4}^C(x)} \right],$$

where $A_{n,x,a} = n^2(C_n \left(a^{-x} - a^{-t} \right)^4)(x)$. Finally using Lemma 1.16, we get the required result. ∎

Remark 1.3 The convergence of the operators C_n in the above theorem takes place for n sufficiently large. Using Lemma 1.14, for $\theta = -1, -2, -3, -4$ and making use of mathematical software, we obtain that

$$\lim_{n\to\infty} A_{n,x,a}$$

$$= \lim_{n\to\infty} n^2 \Big[(C_n a^{-4t})(x) - 4a^{-x}(C_n a^{-3t})(x) + 6a^{-2x}(C_n a^{-2t})(x)$$

$$-4a^{-3x}(C_n a^{-t})(x) + a^{-4x} \Big]$$

$$= \lim_{n\to\infty} n^2 \Bigg[\left(1 + \frac{8x \ln a}{n}\right)^{-n/2} - 4a^{-x}\left(1 + \frac{6x \ln a}{n}\right)^{-n/2}$$

$$+6a^{-2x}\left(1 + \frac{4x \ln a}{n}\right)^{-n/2} - 4a^{-3x}\left(1 + \frac{2x \ln a}{n}\right)^{-n/2} + a^{-4x} \Bigg]$$

$$= 12x^4 a^{-4x}(\ln a)^4.$$

Using the idea of Ditzian [42], we consider the modulus of continuity defined by

$$\omega_1(f, \delta, B, a) = \sup_{|h|\leq\delta, 0 < x < \infty} |f(x) - f(x+h)| a^{-Bx}.$$

The spaces $Lip(\beta, B, a), 0 < \beta \leq 1$ consist of all functions such that $\omega_1(f, \delta, B, a) \leq M\delta^\beta$ for all $\delta < 1$.

Theorem 1.25 *Let $C_n : E \to C(0, \infty)$, where E is the space of functions f with a^{Bt} growth. If $f \in C^2(0, \infty) \cap E$ and $f'' \in Lip(\beta, B, a), 0 < \beta \leq 1$, then for $n > 2Bx \ln a$ and $x \in (0, \infty)$, we have*

$$\left| (C_n f)(x) - f(x) - \frac{x^2}{n} f''(x) \right|$$

$$\leq \left[a^{2Bx} + \frac{C(B, a, x)}{2} + \frac{\sqrt{C(2B, a, x)}}{2} \right] \cdot \frac{2x^2}{n} \cdot \omega_1\left(f'', \sqrt{\frac{6x^2}{n} + \frac{24x^2}{n^2}}, B, a \right),$$

where $C(B, a, x) = a^{2Bx}\left(1 + a^{Bx}\right)$.

Proof For the function $f \in C^2(0, \infty)$, using Taylor's expansion as in Theorem 1.24, and applying Lemma 1.15, we have

$$\left| (C_n f)(x) - f(x) - \frac{x^2}{n} f''(x) \right| \leq \left(C_n \left| \frac{f''(\eta) - f''(x)}{2}(t - x)^2 \right| \right)(x), \quad (1.5.2)$$

where η lies between t and x. In order to complete the proof of the theorem we estimate $\left(C_n \left| \frac{f''(\eta)-f''(x)}{2}(t-x)^2 \right| \right)(x)$. Following [97, pp.101], we have

$$\left(C_n \left| \frac{f''(\eta)-f''(x)}{2}(t-x)^2 \right| \right)(x) \tag{1.5.3}$$

$$\leq \frac{\omega_1\left(f'', h, B, a\right)}{2}\left[\left(C_n\left(a^{2Bx}+a^{Bt}\right)\cdot\left(|t-x|^2+\frac{|t-x|^3}{h}\right)\right)(x)\right].$$

Using Lemma 1.14, we derive that

$$(C_n(t-x)^2 a^{Bt})(x)$$

$$= x^2\left(\frac{n+2}{n}\right)\left(1-\frac{2Bx\ln a}{n}\right)^{-\frac{n}{2}-2} - 2x^2\left(1-\frac{2Bx\ln a}{n}\right)^{-\frac{n}{2}-1}$$

$$+x^2\left(1-\frac{2Bx\ln a}{n}\right)^{-\frac{n}{2}}$$

$$= \frac{x^2}{\left(1-\frac{2Bx\ln a}{n}\right)^{\frac{n}{2}+2}}\left[1+\frac{2}{n}-2\left(1-\frac{2Bx\ln a}{n}\right)+\left(1-\frac{2Bx\ln a}{n}\right)^2\right]$$

$$= \frac{1}{\left(1-\frac{2Bx\ln a}{n}\right)^{\frac{n}{2}+2}}\left[1+\frac{2B^2x^2(\ln a)^2}{n}\right]\frac{2x^2}{n}.$$

Thus, for $n > 2Bx\ln a$, we obtain

$$\left(C_n(t-x)^2 a^{Bt}\right)(x) \leq a^{2Bx}\left(1+a^{Bx}\right)\mu_{n,2}^C(x)$$

$$:= C(a, B, x)\mu_{n,2}^C(x). \tag{1.5.4}$$

Additionally, by the Cauchy-Schwarz inequality we get

$$\left(C_n|t-x|^3 a^{Bt}\right)(x) \leq \sqrt{(C_n(t-x)^2 a^{2Bt})(x)}\cdot\sqrt{\mu_{n,4}^C(x)}$$

$$\leq \sqrt{C(2B, a, x)\mu_{n,2}^C(x)}\cdot\sqrt{\mu_{n,4}^C(x)}. \tag{1.5.5}$$

Substituting $h := \sqrt{\dfrac{\mu_{n,4}^C(x)}{\mu_{n,2}^C(x)}}$ in (1.5.3) and combining (1.5.2), (1.5.4) and (1.5.5), we derive that

$$\left(C_n \left| \frac{f''(\eta) - f''(x)}{2}(t-x)^2 \right| \right)(x)$$

$$\leq \frac{1}{2}\omega_1\left(f'', h, B, a\right)\left[\left(C_n\left(a^{2Bx} + a^{Bt}\right)\cdot\left(|t-x|^2 + \frac{|t-x|^3}{h}\right)\right)(x)\right]$$

$$= \frac{1}{2}\omega_1\left(f'', \sqrt{\frac{\mu_{n,4}^C(x)}{\mu_{n,2}^C(x)}}, B, a\right)\left[2a^{2Bx} + C(B, a, x) + \sqrt{C(2B, a, x)}\right]\mu_{n,2}^C(x).$$

Finally using Lemma 1.16, we get the desired result. ∎

The following lemma is required to study simultaneous approximation.

Lemma 1.17 *There exist the polynomials* $q_{i,j,m}(x)$ *independent of n and t such that*

$$[2x^2]^m \frac{\partial^m}{\partial x^m}[\phi_n^C(x,t)] = \sum_{\substack{2i+j\leq m \\ i,j\geq 0}} n^{i+j}(t-x)^j q_{i,j,m}(x)[\phi_n^C(x,t)].$$

Proof In order to prove the result, it is sufficient to show that

$$\frac{\partial^m}{\partial x^m}\left[x^{-n/2}\exp\left(\frac{-nt}{2x}\right)\right]$$

$$= \sum_{\substack{2i+j\leq m \\ i,j\geq 0}} n^{i+j}(t-x)^j q_{i,j,m}(x)[2x^2]^{-m}\left[x^{-n/2}\exp\left(\frac{-nt}{2x}\right)\right].$$

We shall prove the desired result by applying the principle of mathematical induction. For $m = 1$, we obviously have

$$\frac{\partial}{\partial x}\left[x^{-n/2}\exp\left(\frac{-nt}{2x}\right)\right] = n(t-x)[2x^2]^{-1}\left[x^{-n/2}\exp\left(\frac{-nt}{2x}\right)\right].$$

Thus the result is true for $m = 1$, as in this case $i = 0$, $j = 1$, $q_{i,j,m}(x) = 1$. Let the result be true for m, then

$$\frac{\partial^{m+1}}{\partial x^{m+1}}\left[x^{-n/2}\exp\left(\frac{-nt}{2x}\right)\right]$$

$$= \frac{\partial}{\partial x}\sum_{\substack{2i+j\leq m \\ i,j\geq 0}} n^{i+j}(t-x)^j q_{i,j,m}(x)2^{-m}x^{-2m}x^{-n/2}\exp\left(\frac{-nt}{2x}\right)$$

$$= \sum_{\substack{2i+j\leq m \\ i\geq 0,j\geq 1}} n^{i+j}(t-x)^{j-1}(-jq_{i,j,m}(x))2^{-m}x^{-2m}x^{-n/2}\exp\left(\frac{-nt}{2x}\right)$$

$$+ \sum_{\substack{2i+j\leq m \\ i,j\geq 0}} n^{i+j}(t-x)^j q'_{i,j,m}(x)2^{-m}x^{-2m}x^{-n/2}\exp\left(\frac{-nt}{2x}\right)$$

$$+ \sum_{\substack{2i+j\leq m \\ i,j\geq 0}} n^{i+j}(t-x)^j(-2mq_{i,j,m}(x))2^{-m}x^{-2m-1}x^{-n/2}\exp\left(\frac{-nt}{2x}\right)$$

$$+ \sum_{\substack{2i+j\leq m \\ i,j\geq 0}} n^{i+j+1}(t-x)^j(-q_{i,j,m}(x))2^{-m-1}x^{-2m}x^{-n/2-1}\exp\left(\frac{-nt}{2x}\right)$$

$$+ \sum_{\substack{2i+j\leq m \\ i,j\geq 0}} n^{i+j+1}(t-x)^j(tx^{-2}q_{i,j,m}(x))2^{-m-1}x^{-2m}x^{-n/2}\exp\left(\frac{-nt}{2x}\right).$$

Thus, we have

$$\frac{\partial^{m+1}}{\partial x^{m+1}}\left[x^{-n/2}\exp\left(\frac{-nt}{2x}\right)\right]$$

$$= \sum_{\substack{2(i-1)+(j+1)\leq m \\ i\geq 0,j\geq 1}} n^{i+j}(t-x)^j(-(j+1)q_{i-1,j+1,m}(x))2^{-m}x^{-2m}x^{-n/2}\exp\left(\frac{-nt}{2x}\right)$$

$$+ \sum_{\substack{2i+j\leq m \\ i,j\geq 0}} n^{i+j}(t-x)^j q'_{i,j,m}(x)2^{-m}x^{-2m}x^{-n/2}\exp\left(\frac{-nt}{2x}\right)$$

$$+ \sum_{\substack{2i+j\leq m \\ i,j\geq 0}} n^{i+j}(t-x)^j(-2mq_{i,j,m}(x))2^{-m}x^{-2m-1}x^{-n/2}\exp\left(\frac{-nt}{2x}\right)$$

$$+ \sum_{\substack{2i+(j-1)\leq m \\ i,j\geq 0}} n^{i+j}(t-x)^j(q_{i,j-1,m}(x))2^{-m-1}x^{-2(m+1)}x^{-n/2}\exp\left(\frac{-nt}{2x}\right).$$

This expression has the required form, where

$$q_{i,j,m+1}(x) = -2x^2(j+1)q_{i-1,j+1,m}(x) + 2x^2 q'_{i,j,m}(x)$$
$$- 4mx q_{i,j,m}(x) + q_{i,j-1,m}(x)$$

with $2i + j \le (m+1)$; $i, j \ge 0$ and with the convention that $q_{i,j,m}(x) = 0$ if any one of the constraints is violated. Thus the result holds for $m+1$. This completes the proof of the lemma. ∎

Theorem 1.26 *Let $f \in C(0,\infty)$ with $f(t) = O(a^{\beta t})$, $\beta > 0$, if $f^{(m+2)}$ exists at a fixed point $x \in (0,\infty)$, then we have*

$$\lim_{n\to\infty} n\left[(C_n^{(m)}f)(x) - f^{(m)}(x)\right] = m(m-1)f^{(m)}(x)$$
$$+2mxf^{(m+1)}(x) + x^2 f^{(m+2)}(x).$$

The proof of the above theorem follows using Lemmas 1.15, 1.17 and 1.16, and thus we omit the details.

Corollary 1.4 *Let $f \in C(0,\infty)$ with $f(t) = O(a^{\beta t})$, $\beta > 0$. If f'' exists at a fixed point $x \in (0,\infty)$, then we have*

$$\lim_{n\to\infty} n[(C_n f)(x) - f(x)] = x^2 f''(x).$$

Corollary 1.5 *Let $f \in C(0,\infty)$ with $f(t) = O(a^{\beta t})$, $\beta > 0$. If f''' exists at a fixed point $x \in (0,\infty)$, then we have*

$$\lim_{n\to\infty} n[(C_n' f)(x) - f'(x)] = 2xf''(x) + x^2 f'''(x).$$

1.6   Post-Widder Operators Preserving Exponential Function

The Post–Widder operator (1.1.8) is usually represented in alternative form by

$$(P_n f)(x) = \frac{x^{-n}}{\Gamma(n)} \int_0^\infty e^{-u/x} u^{n-1} f\left(\frac{u}{n}\right) du$$

Müller's Gamma operator are defined as

$$(G_n f)(x) = \frac{x^{n+1}}{\Gamma(n+1)} \int_0^\infty e^{-ux} u^n f\left(\frac{n}{u}\right) du.$$

A more natural form of Müller's Gamma operator is defined as

$$(\tilde{G}_n f)(x) = \frac{x^n}{\Gamma(n)} \int_0^\infty e^{-ux} u^{n-1} f\left(\frac{n}{u}\right) du.$$

Obviously the operators $(\tilde{G}_n f)$ are not of exponential type. One can see that $(P_n e_r)(x) = \frac{x^r \Gamma(n+r)}{n^r \Gamma(n)}$ and $(\tilde{G}_n e_r f)(x) = \frac{n^r x^r \Gamma(n-r)}{\Gamma(n)}$. Some direct estimates on certain general forms of Gamma operators are recently studied in [65].

The Post-Widder operators in slightly modified form are defined for $f \in C[0, \infty)$ as (see [95]):

$$(\hat{P}_n f)(x) := \int_0^\infty \phi_n^{\hat{P}}(x, t) f(t) dt$$

$$= \frac{1}{n!} \left(\frac{n}{x}\right)^{n+1} \int_0^\infty t^n e^{-\frac{nt}{x}} f(t) \, dt.$$

The operators $(\hat{P}_n f)$ as such are different from the Post-Widder operators $(P_n f)$ defined by (1.1.8) and considered by May [118]. These operators $(\hat{P}_n f)$ are not exponential type operators. These operators satisfy the partial differential equation

$$\frac{\partial}{\partial x} \phi_{n,k}^{\hat{P}}(x) = \frac{nt - (n+1)x}{x^2} \phi_{n,k}^{\hat{P}}(x),$$

which is not exactly the required condition for an operator to be an exponential type operator. Also, because of this fact these operators preserve only constant functions.

It was observed by Gupta and Tachev [95] that only two preservations can be made at a time at most, either constant and e_1 or constant and the function e_r, $r > 1, r \in \mathbb{N}$. They [95] dealt with the modification of Post-Widder operators which preserve constants and the test function $e_r, r \in \mathbb{N}$. In another very recent paper, Gupta and Tachev [96] studied a modification of the operators $(\hat{P}_n f)$ preserving exponential functions. Obviously, we have

$$(\hat{P}_n e^{\theta t})(x) = \left(1 - \frac{x\theta}{n}\right)^{-(n+1)}. \tag{1.6.1}$$

Thus the modified operators $\tilde{P}_n$ preserving exponential functions were considered as

$$(\tilde{P}_n f)(x) := \frac{1}{n!} \left[\frac{A}{(1 - e^{-Ax/(n+1)})}\right]^{(n+1)}$$

$$\int_0^\infty t^n e^{-\frac{At}{(1 - e^{-Ax/(n+1)})}} f(t) \, dt, \tag{1.6.2}$$

with $x \in (0, \infty)$ and $(\tilde{P}_n f)(0) = f(0)$, which preserve constants and the test function e^{Ax}.

The following basic lemmas were presented:

Lemma 1.18 ([96]) *For $\theta > 0$, we have that*

$$(\tilde{P}_n e^{\theta t})(x) = \left(1 - \frac{(1 - e^{-Ax/(n+1)})\theta}{A}\right)^{-(n+1)}.$$

Using Lemma 1.18 and denoting $\mu_r^{\tilde{P}_n}(x) = (\tilde{P}_n e_r)(x)$, where $e_r(t) = t^r$, $r \in N \cup \{0\}$, the moments are given by

$$\mu_0^{\tilde{P}_n}(x) = 1,$$

$$\mu_1^{\tilde{P}_n}(x) = \frac{(n+1)}{A}(1 - e^{-Ax/(n+1)}),$$

$$\mu_2^{\tilde{P}_n}(x) = \frac{(n+1)(n+2)}{A^2}(1 - e^{-Ax/(n+1)})^2.$$

Lemma 1.19 ([96]) *The moments of arbitrary order satisfy the following*

$$\mu_k^{\tilde{P}_n}(x) = \frac{(n+1)_k}{A^k}(1 - e^{-Ax/(n+1)})^k, \quad k = 0, 1, \ldots,$$

where we make use of the Pochhammer symbol defined by

$$(c)_0 = 1, \quad (c)_k = c(c+1)\cdots(c+k-1).$$

Lemma 1.20 ([96]) *The central moments $U_r^{\tilde{P}_n}(x) = \tilde{P}_n((t-x)^r, x)$ are given below:*

$$U_k^{\tilde{P}_n}(x) = \sum_{j=0}^{k}(-1)^{k-j}\binom{k}{j}x^{k-j}(1 - e^{-Ax/(n+1)})^j \frac{(n+1)_j}{A^j}, \quad k = 0, 1, \ldots.$$

Also, for each $n \in N$, we have

$$U_1^{\tilde{P}_n}(x) = \frac{(n+1)}{A}(1 - e^{-Ax/(n+1)} - 1) - x,$$

$$U_2^{\tilde{P}_n}(x) = \frac{(n+1)(n+2)}{A^2}(1 - e^{-Ax/(n+1)})^2 + x^2 - 2x\frac{(n+1)}{A}(1 - e^{-Ax/(n+1)}).$$

Lemma 1.21 ([96]) *For the central moments* $U_{2k}^{\widetilde{P}_n}(x) = \widetilde{P}_n((t-x)^{2k}, x)$, *we have*

$$U_{2k}^{\widetilde{P}_n}(x) = O(n^{-k}), n \to \infty, k = 1, 2, 3, \ldots$$

Set $\phi(x) = 1 + e^{Ax}, x \in R^+$ and consider the following weighted spaces:

$$B_\phi(R^+) = \{f : R^+ \to R : |f(x)| \leq C_1(1 + e^{Ax})\},$$
$$C_\phi(R^+) = B_\phi(R^+) \cap C(R^+),$$
$$C_\phi^k(R^+) = \left\{f \in C_\phi(R^+) : \lim_{x \to \infty} \frac{f(x)}{1 + e^{Ax}} = C_2 < \infty\right\},$$

where C_1, C_2 are constants depending on f. The norm is defined as

$$\|f\|_\phi = \sup_{x \in R^+} \frac{|f(x)|}{1 + e^{Ax}}.$$

Based on the above lemmas, the following theorems were estimated for the modified Post-Widder operators.

Theorem 1.27 ([96]) *For each* $f \in C_\phi^k(R^+)$, *we have*

$$\lim_{n \to \infty} \|\widetilde{P}_n f - f\|_\phi = 0.$$

Also the following quantitative asymptotic formula was established.

Theorem 1.28 ([96]) *Let* $\widetilde{P}_n : E \to C[0, \infty)$ *be sequence of linear positive operators of Post-Widder type defined in (1.6.2). Then*

$$|(\widetilde{P}_n f)(x) - f(x)| \leq e^{Ax}[3 + C(n, x)]\omega_1\left(f, \sqrt{U_2^{\widetilde{P}_n}(x)}, A\right),$$

where

$$C(n, x) = 2 \sum_{k=1}^{\infty} \frac{A^k}{k!} \sqrt{U_{2k}^{\widetilde{P}_n}(x)}, \quad n \to \infty$$

for fixed $x \in [0, \infty)$ *and the first order exponential modulus of continuity, studied by Ditzian in [42] and defined as*

$$\omega_1(f, \delta, A) := \sup_{h \leq \delta, 0 \leq x < \infty} |f(x) - f(x+h)|e^{-Ax}.$$

1.7 Semi-Exponential Operators

In 2005, Tyliba and Wachnicki [155] extended the results of Ismail and May [103], to propose a more general family of operators, in which they considered the operators L_λ^β such that $L_\lambda^\beta(e_1, x) \neq e_1(x)$. They introduced some operators, which are not exponential, but similar to exponential type operators. The general form of these operators takes the following form:

$$\left(L_\lambda^\beta f\right)(x) = \sum_k \psi_{\lambda,k}^{L^\beta}(x) f\left(\frac{k}{\lambda}\right) \quad \text{or} \quad \int_{-\infty}^{\infty} \phi_\lambda^{L^\beta}(x, t) f(t) dt, \qquad (1.7.1)$$

whose kernels satisfy the partial differential equations

$$\frac{\partial}{\partial x} \psi_{\lambda,k}^{L^\beta}(x) = \frac{k - \lambda x}{p(x)} \psi_{\lambda,k}^{L^\beta}(x) - \beta \psi_{\lambda,k}^{L^\beta}(x) \qquad (1.7.2)$$

$$\frac{\partial}{\partial x} \phi_\lambda^{L^\beta}(x, t) = \frac{\lambda(t - x)}{p(x)} \phi_\lambda^{L^\beta}(x, t) - \beta \phi_\lambda^{L^\beta}(x, t), \qquad (1.7.3)$$

respectively. Here β is a non-negative real number and the other normalization conditions are the same as in (1.1.1). Recently, Herzog [99] also studied the operators (1.7.1) and termed them as semi-exponential type operators.

Remark 1.4 [155, Lemma 2.1] For the semi-exponential operators $(L_\lambda^\beta f)$, we have

$$(L_\lambda^\beta e_0)(x) = 1$$

$$(L_\lambda^\beta e_1)(x) = e_1(x) + \frac{\beta p(x)}{\lambda}$$

$$(L_\lambda^\beta e_2)(x) = e_2(x) + \frac{(1 + 2\beta x) p(x)}{\lambda} + \frac{\beta p(x)(p'(x) + \beta p(x))}{\lambda^2}.$$

The semi-exponential Gauss-Weierstrass operator, for $x \in (-\infty, \infty)$ and $p(x) = 1$, considered in [155] is defined as:

$$\left(W_\lambda^\beta f\right)(x) = \sqrt{\frac{\lambda}{2\pi}} \int_{-\infty}^{\infty} \exp\left(\frac{-\lambda(t - x - \frac{\beta}{\lambda})^2}{2}\right) f(t) dt. \qquad (1.7.4)$$

Additionally, the semi-exponential Szász-Mirakyan operator, for $x \in [0, \infty)$ and $p(x) = x$, obtained in [155] is defined by

$$\left(S_\lambda^\beta f\right)(x) = \sum_{k=0}^{\infty} e^{-(\lambda+\beta)x} \frac{(\lambda + \beta)^k x^k}{k!} f\left(\frac{k}{n}\right). \qquad (1.7.5)$$

The semi-exponential Post-Widder operators for $x \in (0, \infty)$, $p(x) = x^2$ is defined by

$$(P_\lambda^\beta f)(x) = \frac{\lambda}{x^\lambda \exp(\beta x)} \int_0^\infty \frac{\left(\frac{\lambda t}{\beta}\right)^{(\lambda-1)/2} I_{\lambda-1}(2\sqrt{\lambda \beta t})}{\exp(\lambda t/x)} f(t)dt, \quad (1.7.6)$$

where $I_{\lambda-1}$ is the modified Bessel function of the first kind. A probabilistic approach for $(P_\lambda^\beta f)$ is also demonstrated by Herzog [99]. Based on these recent studies one can also establish several other examples of semi-exponential type operators by using the Laplace transform.

Chapter 2
Modifications of Certain Operators

2.1 Introduction

The first integral variant of Bernstein polynomials was introduced by Kantorovich in [110]. Later, in the year 1967, Durrmeyer [48] proposed yet another integral modification of the classical Bernstein polynomials in order to approximate integrable functions. After a long period, about four decades ago, in the year 1981, Derriennic [41] studied these operators in detail and estimated some approximation results. It was shown by Derriennic [41] that the operators of Bernstein–Durrmeyer are positive contractions in L_p, are self-adjoint and satisfy the commutative property. In 1989, Ditzian and Ivanov [45] provided some results on ordinary and simultaneous approximation in terms of weighted modulus of smoothness. Later, Heilmann and Müller [100] proposed a unified approach to introduce Durrmeyer type operators, which also include the Baskakov and Szász Durrmeyer operators. Several researchers worked in this direction and proposed many usual and hybrid operators of Durrmeyer type. A thorough presentation of moments for several operators was recently provided by the authors in [89]. Additionally, for other approximation properties, we refer the readers to [91].

2.2 Kantorovich Operators

The Bernstein–Kantorovich operators are defined by

$$(K_n^B f)(x) = (n+1) \sum_{k=0}^{n} p_{n,k}(x) \int_{k/(n+1)}^{(k+1)/(n+1)} f(t)dt, \quad x \in [0, 1], \quad (2.2.1)$$

where $p_{n,k}(x)$ is the Bernstein basis given in (1.1.5).

© The Author(s), under exclusive license to Springer Nature Switzerland AG 2021
V. Gupta, M. T. Rassias, *Computation and Approximation*, SpringerBriefs
in Mathematics, https://doi.org/10.1007/978-3-030-85563-5_2

Theorem 2.1 *If the differential operator is denoted by D, then the Bernstein polynomials B_{n+1} of order $n+1$ (see (1.1.5)) are connected with the Bernstein–Kantorovich (2.2.1) by the relation:*

$$(K_n^B f)(x) = (D \circ B_{n+1} \circ F)(x),$$

where F is the antiderivative operator of f given by

$$F(x) = \int_0^x f(t)dt.$$

Proof To prove this relation, we use the identity

$$\left(p_{n,k}(x)\right)' = n\left[p_{n-1,k-1}(x) - p_{n-1,k}(x)\right], \quad 0 < k < n.$$

We have

$$(D \circ B_{n+1} \circ F)(x) = D\left((B_{n+1}F)(x)\right) = D\left(\sum_{k=0}^{n+1} p_{n+1,k}(x) F\left(\frac{k}{n+1}\right)\right)$$

$$= \sum_{k=0}^{n} p_{n+1,k}'(x) F\left(\frac{k}{n+1}\right)$$

$$= (n+1)\sum_{k=0}^{n}\left[p_{n,k-1}(x) - p_{n,k}(x)\right] F\left(\frac{k}{n+1}\right)$$

$$= (n+1)\sum_{k=0}^{n} p_{n,k}(x)\left(F\left(\frac{k+1}{n+1}\right) - F\left(\frac{k}{n+1}\right)\right)$$

$$= (n+1)\sum_{k=0}^{n} p_{n,k}(x) \int_{k/n+1}^{(k+1)/(n+1)} f(t)dt$$

$$= (K_n^B f)(x).$$

∎

In 1983, Nagel [133] introduced the Kantorovich operators of second order as follows:

$$(Q_n^B f)(x) = \frac{d^2}{dx^2}(B_{n+2}\mathsf{F})(x),$$

where

$$\mathsf{F} = \int_0^x F(t)dt - x\int_0^1 F(t)dt, \; F(t) = \int_0^t f(w)dw.$$

For a convex operator L of order $k - 1$, and thus satisfying

$$L\left(\prod_{k-1}\right) \subset \prod_{k-1},$$

consider $I_k : C[0, 1] \to C[0, 1]$ given by
$I_k f = f$, if $k = 0$ and

$$(I_k f)(x) = \int_0^x \frac{(x - t)^{k-1}}{(k - 1)!} f(t) dt \quad \text{if } k \geq 1.$$

The k-th order Kantorovich modification of L proposed by Gonska et al. in [59] is given by

$$Q^k := D^k \circ L \circ I_k,$$

where $D^k = \frac{d^k}{dx^k}$.

Also, the Baskakov–Kantorovich operators considered by Abel et al. [3, (1)] are defined by

$$(K_n^V f)(x) = n \sum_{k=0}^{\infty} v_{n+1,k}(x) \int_{k/n}^{(k+1)/n} f(t) dt, \qquad (2.2.2)$$

where $v_{n,k}(x)$ is the Baskakov basis function defined in (1.1.6).

Theorem 2.2 *The Baskakov–Kantorovich operators K_n^V are connected with the Baskakov operators V_n (see (1.1.6)) by the following relation:*

$$(K_n^V f)(x) = (D \circ V_n \circ F)(x),$$

where F is defined in Theorem 2.1.

Proof To prove this relation, we use the identity

$$\left(v_{n,k}(x)\right)' = n\left[v_{n+1,k-1}(x) - v_{n+1,k}(x)\right].$$

We have

$$(D \circ V_n \circ F)(x) = \sum_{k=0}^{\infty} v'_{n,k}(x) F\left(\frac{k}{n}\right)$$

$$= n \sum_{k=0}^{\infty} \left[v_{n+1,k-1}(x) - v_{n+1,k}(x)\right] F\left(\frac{k}{n}\right)$$

$$= n \sum_{k=0}^{\infty} v_{n+1,k}(x) \left(F\left(\frac{k+1}{n}\right) - F\left(\frac{k}{n}\right) \right)$$

$$= n \sum_{k=0}^{\infty} v_{n+1,k}(x) \int_{k/n}^{(k+1)/n} f(t)dt$$

$$= (K_n^V f)(x).$$

$\blacksquare$

Remark 2.1 Another form of the Baskakov–Kantorovich operators considered by Xhang and Zhu [159] is defined by

$$(K_n^{\hat{V}} f)(x) = n \sum_{k=0}^{\infty} v_{n,k}(x) \int_{k/n}^{(k+1)/n} f(t)dt. \qquad (2.2.3)$$

If we consider

$$F_n(x) = \int_0^x f_n(t)dt, \quad f_n(t) = f\left(\frac{(n-1)t}{n}\right),$$

then we have the following connection between the Baskakov operators (1.1.6) and the Baskakov–Kantorovich operators (2.2.3) (see [159]):

$$(K_n^{\hat{V}} f)(x) = (D \circ V_{n-1} \circ F_n)(x),$$

where V_{n-1} is the Baskakov operators (see (1.1.6)).

Additionally, the Szász–Mirakyan–Kantorovich operators are defined by

$$(K_n^S f)(x) = n \sum_{k=0}^{\infty} s_k(nx) \int_{k/n}^{(k+1)/n} f(t)dt, \quad x \in [0, \infty), \qquad (2.2.4)$$

where $s_k(nx)$ is the Szász–Mirakyan basis function defined in (1.1.7).

Theorem 2.3 *We have the following connection between the Szász–Mirakyan operators (1.1.7) and the Szász–Mirakyan–Kantorovich operators (2.2.4):*

$$(K_n^S f)(x) = (D \circ S_n \circ F)(x),$$

where F is defined in Theorem 2.1.

Proof To prove this relation, we use the identity

$$\left(s_{n,k}(x)\right)' = n \left[s_{n,k-1}(x) - s_{n,k}(x)\right].$$

We have

$$
(D \circ S_n \circ F)(x) = \sum_{k=0}^{\infty} s'_{n,k}(x) F\left(\frac{k}{n}\right)
$$

$$
= n \sum_{k=0}^{\infty} \left[s_{n,k-1}(x) - s_{n,k}(x) \right] F\left(\frac{k}{n}\right)
$$

$$
= n \sum_{k=0}^{\infty} s_{n,k}(x) \left(F\left(\frac{k+1}{n}\right) - F\left(\frac{k}{n}\right) \right)
$$

$$
= n \sum_{k=0}^{\infty} s_{n,k}(x) \int_{k/n}^{(k+1)/n} f(t)dt
$$

$$
= (K_n^S f)(x).
$$

∎

The operator $M_n, n \in N$, of Meyer–König–Zeller (MKZ) associated with the bounded function $f : [0, 1] \to R$, the so-called Bernstein power series is defined by

$$
(M_n f)(x) = \sum_{k=0}^{\infty} m_{n,k}(x) f\left(\frac{k}{n+k}\right),
$$

where

$$
m_{n,k}(x) = \binom{n+k}{k} x^k (1 - x)^{n+1},
$$

converges for $x \in [0, 1)$. Müller [132] obtained a connection of the MKZ operators with its Kantorovich variant as follows:

$$
D(M_n f)(x) = (n + 1)(1 - x)^n \sum_{k=0}^{\infty} \binom{n+k+1}{k} x^k
$$

$$
\times \left[f\left(\frac{k+1}{n+k+1}\right) - f\left(\frac{k}{n+k}\right) \right],
$$

$0 \le x < 1$. If $f \in L_p[0, 1]$ and I is the antiderivative of the function f, then for $0 \le x < 1$, the Kantorovich variant is given by

$$
(K_n^M f)(x) = \sum_{k=0}^{\infty} \hat{m}_{n,k}(x) \int_{k/n+k}^{(k+1)/n+k+1} f(t)dt, \quad x \in [0, 1), \qquad (2.2.5)
$$

where

$$\hat{m}_{n,k}(x) = (n+1)\binom{n+k+1}{k}x^k(1-x)^n.$$

Furthermore, the following connection was established in [132] between the MKZ operators and its Kantorovich variant:

$$(K_n^M f)(x) = (D \circ M_n \circ I)(x).$$

Another interesting sequence of linear positive operators in approximation defined on the space $R^{[0,\infty)}$ of real functions on the interval $[0, \infty)$ are the well-known Bleimann, Butzer and Hahn (BBH) operators [30], which are defined as

$$(H_n f)(x) = \frac{1}{(1+x)^n} \sum_{k=0}^{n} \binom{n}{k} x^k f\left(\frac{k}{n+1-k}\right), \quad x \in [0, \infty).$$

The Bleimann, Butzer and Hahn operators are closely connected to the Bernstein operators (see [12] and [106]). Abel and Ivan [6] considered the operators $U :$ $R^{[0,\infty)} \to R^{[0,1]}$

$$U(f, t) := \begin{cases} (1-t)f\left(\frac{t}{1-t}\right), & : t \in [0, 1) \\ 0, & : t = 1 \end{cases}$$

and $V : R^{[0,1)} \to R^{[0,\infty)}$

$$V(g, x) := (1+x)g\left(\frac{x}{1+x}\right), \quad x \in [0, \infty).$$

Using the above notation, the following relation was established (see [12] and [106]):

$$H_n = V \circ B_{n+1} \circ U, n = 1, 2, 3, \ldots.$$

The Kantorovich variant of BBH operators H_n given in [158] is defined as

$$(K_n^H f)(x) = \frac{n+2}{(1+x)^n} \sum_{k=0}^{n} \binom{n}{k} x^k \int_{(k+1)/(n+1-k)}^{k/(n+2-k)} \frac{f(t)}{(1+t)^2} dt. \quad (2.2.6)$$

It was observed in [158] that direct application of the operators U and V to the classical Bernstein–Kantorovich operators $(K_n^B f)$ yields the operator

$$K_n^* = V \circ K_n^B \circ U$$

given by

$$(K_n^* f)(x) = \frac{n+2}{(1+x)^n} \sum_{k=0}^{n+1} \binom{n+1}{k} x^k \int_{(k+1)/(n+1-k)}^{k/(n+2-k)} \frac{f(t)}{(1+t)^3} dt. \quad (2.2.7)$$

This operator does not preserve constant functions since

$$(K_n^* e_0)(x) = 1 + \frac{(x-1)}{2(n+2)}.$$

Moreover, it requires an unpleasant growth condition on f ensuring the convergence of the integral for $k = n + 1$. If $(1 + t)^{-1}$ in the last integral is replaced by the constant $1 - \frac{k}{n+1}$, then the last summand vanishes and the authors of [158] obtained the operators K_n^H as defined in (2.2.6). This was the main reason that the authors of [158] considered the operators (2.2.6) as the Kantorovich variants of BBH operators.

There are many operators of discrete type available in the literature, which do not satisfy the partial differential equation indicated in (1.1.1). Here, we mention some of those operators for reference purposes and for the convenience of the readers.

The well-known mathematician Alexandru Lupaş in [114] proposed for $x \geq 0$ the following important operators:

$$(L_n f)(x) = \sum_{k=0}^{\infty} l_k(nx) f\left(\frac{k}{n}\right), \quad (2.2.8)$$

where

$$l_k(nx) = \frac{(nx)_k}{2^k \cdot k!} \cdot 2^{-nx},$$

and the rising factorial is given by $(nx)_m = \prod_{i=0}^{m-1}(nx + i), m \geq 1; \ (nx)_0 = 1$. These operators are not as such exponential type operators. Agratini [13] proposed the Kantorovich variant of the operators (2.2.8) as

$$(K_n^L f)(x) = n \sum_{k=0}^{\infty} l_k(nx) \int_{k/n}^{(k+1)/n} f(t) dt, \quad x \geq 0. \quad (2.2.9)$$

Very recently, in [62] and the references therein, the connection between (2.2.8) and (2.2.9) was established as follows:

$$(K_n^L f) =: n((1 - e^{-D/n}) \circ L_n \circ F) = n(\nabla \circ L_n \circ F),$$

where $F(x) = \int_0^x f(t) dt$. Also, the j-th order Lupaş-Kantorovich operator is defined as

$$(K_n^{L,j} f) = n^j(\nabla^j \circ L_n \circ I_j), \quad (2.2.10)$$

where

$$I_j(f; x) = \begin{cases} f(x), & j = 0 \\ \displaystyle\int_0^x \frac{(x-t)^{j-1}}{(j-1)!} f(t)dt, & j \geq 1, \end{cases}$$

and ∇ stands for the backward difference operator for the function $f(nx)$ with unit step length.

Jain–Pethe [108] proposed, for $x \geq 0$, the following important discrete operators:

$$(J_n^{[\alpha]} f)(x) = \sum_{k=0}^{\infty} s_{n,k}^{[\alpha]}(x) f\left(\frac{k}{n}\right), \tag{2.2.11}$$

where

$$s_{n,k}^{[\alpha]}(x) = \frac{n^k}{k!} (n\alpha + 1)^{-(k\alpha+x)/\alpha} . x^{(k,-\alpha)}$$

and

$$x^{(k,-\alpha)} = \prod_{r=0}^{k-1} (x + r\alpha), k \geq 1; \quad x^{(0,-\alpha)} = 1.$$

In particular, if $\alpha = 1/n$, we immediately get the Lupaş operators introduced in [114] and later discussed in [13]. These operators in an alternate form are defined by (2.8.3).

The moment generating function (m.g.f.) of $J_n^{[\alpha]} f$ is computed as

$$(J_n^{[\alpha]} e^{\lambda t})(x) = [1 + n\alpha(1 - e^{\lambda/n})]^{-x/\alpha}.$$

The r-th order moment, where $e_r(t) = t^r, r = 0, 1, 2, \ldots$ of (2.2.11), can be obtained from the m.g.f. by the following relation:

$$(J_n^{[\alpha]} e_r)(x) = \left[\frac{\partial^r}{\partial \lambda^r} [1 + n\alpha(1 - e^{\lambda/n})]^{-x/\alpha} \right]_{\lambda=0}. \tag{2.2.12}$$

Thus, by using the concept of m.g.f., one can easily find the moments and can avoid lengthy computations, which have been done in Lemma 2.1 and Proposition 2.1 of [44]. Also, the central moments can be obtained by applying the change of scale property of m.g.f. as follows:

$$(J_n^{[\alpha]} (e_1 - xe_0)^r)(x) = \left[\frac{\partial^r}{\partial \lambda^r} e^{-\lambda x} [1 + n\alpha(1 - e^{\lambda/n})]^{-x/\alpha} \right]_{\lambda=0}.$$

One can also avoid the computation indicated in [44, Remark 2.1].

The Kantorovich type integral variant of the Jain–Pethe operators $J_n^{[\alpha]} f$ (see 2.2.11) was proposed in [44] as

$$(K_n^{[\alpha]} f)(x) = n \sum_{k=0}^{\infty} s_{n,k}^{[\alpha]}(x) \int_{k/n}^{(k+1)/n} f(v) dv, \, x \geq 0. \tag{2.2.13}$$

Theorem 2.4 (see [81]) *Between the Jain-Pethe operators and its Kantorovich variant, we have the following link:*

$$(K^{[\alpha]} f) = (\nabla \circ J_n^{[\alpha]} \circ F_\alpha), \tag{2.2.14}$$

where $F_\alpha(x) = \frac{1}{\alpha} \int_0^x f(v) dv, \alpha \neq 0$ *and* ∇ *is the backward difference operator with step length* α.

Proof Firstly, we observe that

$$\nabla x^{(k,-\alpha)} = k\alpha x^{(k-1,-\alpha)}$$

and

$$\nabla (n\alpha + 1)^{-x/\alpha} = (n\alpha + 1)^{-x/\alpha} - (n\alpha + 1)^{-(x-\alpha)/\alpha} = -n\alpha (n\alpha + 1)^{-x/\alpha}.$$

By the well-known property of the backward difference operators:

$$\nabla (hg) = h(\nabla g) + g(\nabla h) - (\nabla h)(\nabla g),$$

we can write

$$\nabla \left[x^{(k,-\alpha)} \cdot (n\alpha + 1)^{-x/\alpha} \right]$$

$$= -n\alpha x^{(k,-\alpha)} (n\alpha + 1)^{-x/\alpha} + k\alpha x^{(k-1,-\alpha)} (n\alpha + 1)^{-x/\alpha}$$

$$+ nk\alpha^2 x^{(k-1,-\alpha)} (n\alpha + 1)^{-x/\alpha}$$

$$= k\alpha x^{(k-1,-\alpha)} \cdot (n\alpha + 1)^{-(x-\alpha)/\alpha} - n\alpha x^{(k,-\alpha)} (n\alpha + 1)^{-x/\alpha}.$$

Thus, by the definition of $s_{n,k}^{[\alpha]}(x)$ given in (2.2.11), we get

$$\nabla (s_{n,k}^{[\alpha]}(x)) = n\alpha [s_{n,k-1}^{[\alpha]}(x) - s_{n,k}^{[\alpha]}(x)].$$

Using the above identity, we have

$$(\nabla \circ J_n^{[\alpha]} \circ F_\alpha)(x) = \sum_{k=0}^{\infty} (\nabla s_{n,k}^{[\alpha]}(x)) F_\alpha \left(\frac{k}{n} \right)$$

$$= n\alpha \sum_{k=0}^{\infty} [s_{n,k-1}^{[\alpha]}(x) - s_{n,k}^{[\alpha]}(x)] F_\alpha \left(\frac{k}{n} \right)$$

$$= n\alpha \sum_{k=0}^{\infty} s_{n,k}^{[\alpha]}(x) \left[F_\alpha \left(\frac{k+1}{n} \right) - F_\alpha \left(\frac{k}{n} \right) \right]$$

$$= n \sum_{k=0}^{\infty} s_{n,k}^{[\alpha]}(x) \int_{k/n}^{(k+1)/n} f(t)dt = (K_n^{[\alpha]} f)(x).$$

Thus, the proof follows.   ∎

For α a non-negative parameter and $x \in [0, 1]$, the generalization of the well-known Bernstein polynomial was given by Stancu [152] as follows:

$$(P_n^{(\alpha)} f)(x) = \sum_{k=0}^{n} b_{n,k}^{(\alpha)}(x) f \left(\frac{k}{n} \right), \qquad (2.2.15)$$

where

$$b_{n,k}^{(\alpha)}(x) = \binom{n}{k} \frac{x^{[k,-\alpha]}(1-x)^{[n-k,-\alpha]}}{1^{[n,-\alpha]}}$$

and $b_{n,k}^{(\alpha)}(x) = 0$ if $k < 0$ or $k > n$ with

$$x^{[j,-\alpha]} = \prod_{r=0}^{j-1} (x + r\alpha), \, j \geq 1; \quad x^{[0,-\alpha]} = 1.$$

These operators are based on the Pólya distribution, and for $\alpha = 0$ we get the Bernstein polynomials. The operators were studied in detail by Miclăuş [125]. The special case $\alpha = 1/n$ was discussed by Lupaş and Lupaş [117].

The Kantorovich type integral variant of the operators (2.2.15) was introduced by Razi in [146] as follows:

$$(K_n^{(\alpha)} f)(x) = (n+1) \sum_{k=0}^{n} b_{n,k}^{(\alpha)}(x) \int_{k/(n+1)}^{(k+1)/(n+1)} f(t)dt, \qquad (2.2.16)$$

where $b_{n,k}^{(\alpha)}(x)$ is given in (2.2.15). The special case of these operators was considered by Agrawal et al. [15], who obtained some direct local and global results on the operators (2.2.16). Also, as a special case when $\alpha = 0$, we get the usual Bernstein–Kantorovich operators. We point out here that the operators $P_n^{(\alpha)}$ and $K_n^{(\alpha)}$ are not directly connected. We also note that the Kantorovich variant has to be slightly modified to have a connection with discrete operators $P_n^{(\alpha)}$. Such a connection will be provided by the first author in forthcoming papers.

The operators due to Stancu [152] are based on the inverse Pólya–Eggenberger distribution and constitute a generalization of the well-known Baskakov operators. For α a non-negative parameter and $x \in [0, \infty)$, the operators are defined as follows:

$$(V_n^{(\alpha)} f)(x) = \sum_{j=0}^{\infty} v_{n,j}^{(\alpha)}(x) f\left(\frac{j}{n}\right), \tag{2.2.17}$$

where

$$v_{n,j}^{(\alpha)}(x) = \frac{(n)_j}{j!} \frac{1^{[n,-\alpha]} x^{[j,-\alpha]}}{(1+x)^{[n+j,-\alpha]}}$$

with $(n)_j = n(n+1) \cdots (n+j-1)$, $(n)_0 = 1$ and

$$x^{(j,-\alpha)} = \prod_{r=0}^{j-1}(x+r\alpha), \quad j \geq 1; \quad x^{(0,-\alpha)} = 1.$$

In particular, if $\alpha = 0$, we get the Baskakov operators.

The Kantorovich type integral variant of the operators (2.2.17) for $x \in [0, \infty)$ was considered by Deo et al. [39] as follows:

$$(\widetilde{K}_n^{(\alpha)} f)(x) = (n-1) \sum_{j=0}^{n} v_{n,j}^{(\alpha)}(x) \int_{j/(n-1)}^{(j-1)/(n+1)} f(t)dt, \tag{2.2.18}$$

where $v_{n,j}^{(\alpha)}(x)$ is given in (2.2.17).

Additionally, while introducing an operator, there must be some significance into defining the new modification of the operators. We point out here that the operators $V_n^{(\alpha)}$ and $\widetilde{K}_n^{(\alpha)}$ are not directly connected. We can determine a connection in the slightly modified form of one of the two operators in terms of backward differences, which will be presented by the first author and collaborators in the forthcoming research.

The Jain [107] operators are defined by

$$(J_n^{[\beta]} f)(x) = \sum_{k=0}^{\infty} s_k(\beta, nx) f\left(\frac{k}{n}\right), \quad x \geq 0, \tag{2.2.19}$$

where

$$s_k(\beta, nx) = \frac{nx(nx + k\beta)^{k-1}}{k!} e^{-(nx+k\beta)}.$$

The Kantorovich type Jain operators (see [156]) are defined as follows:

$$(\tilde{K}_n^{[\beta]}f)(x) = n \sum_{k=0}^{\infty} s_k(\beta, nx) \int_{k/n}^{(k+1)/n} f(t)dt, \quad x \geq 0. \qquad (2.2.20)$$

Moreover, the operator $(\tilde{K}_n^{[\beta]}f)$ is not directly connected with $(J_n^{[\beta]}f)$. Thus, in order to establish a connection, we must modify the Kantorovich variant as follows:

$$(K_n^{[\beta]}f)(x) = n \sum_{k=0}^{\infty} s_k(\beta, nx + \beta) \int_0^{(k+1)/n} f(t)dt$$

$$-n \sum_{k=0}^{\infty} s_k(\beta, nx) \int_0^{k/n} f(t)dt. \qquad (2.2.21)$$

But the form (2.2.21) is not suitable as far as approximation results are concerned. This happens because the derivatives of the Jain basis function $s_k(\beta, nx)$ are not positive, which is clear from the following theorem:

Theorem 2.5 *Between the Jain operators and its Kantorovich variant, we have the following relation:*

$$(K_n^{[\beta]}f) = (D \circ J_n^{[\beta]} \circ F),$$

where $F(x) = \int_0^x f(t)dt$.

Proof Clearly, we have

$$D\left[(nx).(nx + k\beta)^{k-1}\right] = [(k-1)n^2x(nx + k\beta)^{k-2} + n(nx + k\beta)^{k-1}].$$

Thus,

$$D\left[e^{-(nx+k\beta)}\{(nx).(nx + k\beta)^{k-1}\}\right]$$

$$= e^{-(nx+k\beta)}[(k-1)n^2x(nx + k\beta)^{k-2} + n(nx + k\beta)^{k-1}]$$

$$-(n^2x).(nx + k\beta)^{k-1}e^{-(nx+k\beta)}$$

$$= kn^2xe^{-(nx+k\beta)}(nx + k\beta)^{k-2} - n^2xe^{-(nx+k\beta)}(nx + k\beta)^{k-2}$$

$$+ne^{-(nx+k\beta)}(nx + k\beta)^{k-1} - n^2x(nx + k\beta)^{k-1}e^{-(nx+k\beta)}$$

$$= kn^2xe^{-(nx+k\beta)}(nx + k\beta)^{k-2} + nk\beta e^{-(nx+k\beta)}(nx + k\beta)^{k-2}$$

$$-n^2x(nx + k\beta)^{k-1}e^{-(nx+k\beta)}$$

$$= n[knxe^{-(nx+k\beta)}(nx + k\beta)^{k-2} + k\beta e^{-(nx+k\beta)}(nx + k\beta)^{k-2}$$

$$-nx(nx+k\beta)^{k-1}e^{-(nx+k\beta)}]$$
$$= n[k(nx+\beta)e^{-(nx+k\beta)}(nx+k\beta)^{k-2} - nx(nx+k\beta)^{k-1}e^{-(nx+k\beta)}]$$

implying

$$D(s_k(\beta, nx)) = n[s_{k-1}(\beta, nx+\beta) - s_k(\beta, nx)]. \tag{2.2.22}$$

Using (2.2.22), we have

$$(D \circ J_n^{[\beta]} \circ F)(x)$$

$$= \sum_{k=0}^{\infty} (Ds_k(\beta, nx)) F\left(\frac{k}{n}\right)$$

$$= n \sum_{k=0}^{\infty} [s_{k-1}(\beta, nx+\beta) - s_k(\beta, nx)] F\left(\frac{k}{n}\right)$$

$$= n \sum_{k=0}^{\infty} s_k(\beta, nx+\beta) F\left(\frac{k+1}{n}\right) - n \sum_{k=0}^{\infty} s_k(\beta, nx) F\left(\frac{k}{n}\right)$$

$$= n \sum_{k=0}^{\infty} s_k(\beta, nx+\beta) \int_0^{(k+1)/n} f(t)dt - n \sum_{k=0}^{\infty} s_k(\beta, nx)$$

$$\times \int_0^{k/n} f(t)dt = (K_n^{[\beta]} f)(x).$$

This completes the proof of the theorem.

∎

2.3 Some Operators and Affine Functions

The operators L_n reproduce the affine functions property implied by the following two relations: $L_n e_0 = e_0$ and $L_n e_1 = e_1, n \in N$. It is well known that discrete operators are not suitable for approximating discontinuous functions, and they were generalized into operators of integral type. Here, we discuss the Kantorovich method which leads to the approximation process in the spaces of integrable functions. Agratini in [14] considered the Kantorovich type general sequence of the operators (1.1.1) as

$$(K_n^L f)(x) = \frac{1}{p_n} \sum_k \psi_{n,k}^L \left(x - \frac{p_n}{2}\right) \int_{x_{n,k}}^{x_{n,k+1}} f(t)dt, \tag{2.3.1}$$

where $x_{n,k+1} - x_{n,k} = p_n$ and $\psi_{n,k}^L$ is the basis function of an operator. Agratini in [14] obtained the following moments for the generalized Kantorovich operators:

Remark 2.2 Let $(K_n^L f)$ be defined as in (2.3.1). Then we have

$$(K_n^L e_0)(x) = 1$$
$$(K_n^L e_1)(x) = x$$

$$(K_n^L e_0)(x) = (L_n e_2)\left(x - \frac{p_n}{2}\right) + p_n x - \frac{p_n^2}{6},$$

where L_n is the discrete operator mentioned in (1.1.1).

It was observed by Agratini [14] that such a representation reduces the error over the standard Kantorovich operators for many different cases. He considered such a representation (2.3.1) and provided examples of the modified Szász–Kantorovich, Baskakov–Kantorovich and Stancu–Kantorovich operators.

In 2003, Gupta–Maheshwari [85] proposed the following Durrmeyer type integral modification of the Bernstein polynomials as

$$(\overline{R}_n f)(x) = n \sum_{k=0}^{n} p_{n,k}(x) \int_0^1 p_{n-1,k-1}(t) f(t) dt + (1-x)^n f(0), \quad (2.3.2)$$

where the Bernstein basis function is given by

$$p_{n,k}(x) = \binom{n}{k} x^k (1-x)^{n-k}.$$

It was observed in [85] that the operators $(\overline{R}_n f)$ preserve only constant functions. They obtained the following moments for such operators:

Remark 2.3 Let $(K_n^L f)$ be defined as in (2.3.1). Then, we have

$$(\overline{R}_n e_0)(x) = 1$$
$$(\overline{R}_n e_1)(x) = \frac{nx}{n+1}$$
$$(\overline{R}_n e_2)(x) = \frac{nx(x(n-1)+2)}{(n+1)(n+2)}.$$

In order to preserve the affine functions, Gupta and Duman [82] (see also [78, pp. 144]) considered $q_n(x) = \frac{(n+1)x}{n}$, to obtain the following operators by using the restriction $x \in [0, 0.5]$:

$$(\overline{T}_n f)(x) = n \sum_{k=0}^{n} t_{n,k}(x) \int_0^1 p_{n-1,k-1}(t) f(t) dt$$

$$+ \left(1 - \frac{(n+1)}{n}x\right)^n f(0), \tag{2.3.3}$$

where $x \in [0, 0.5]$, $n \in \mathbb{N}$, the term $p_{n-1,k-1}(t)$ is given in (2.3.2) and

$$t_{n,k}(x) = p_{n,k}(q_n(x)) = \binom{n}{k} \frac{(n+1)^k x^k (n - (n+1)x)^{n-k}}{n^n}.$$

The modified operators (2.3.3) preserve the affine functions. It was pointed out in [82] that considering the operators to preserve affine functions leads to the reduction of error, but sometimes a drawback occurs, i.e. we may get a reduced error in further compact interval [0.4, 0.5], while the modified operators are defined in [0, 0.5]. Some other applications concerning affine functions have been discussed in [78, Ch.5].

Very recently, Bustamante in [31] considered the modification of Szász–Kantorovich operators

$$(\overline{K}_n^S f)(x) = n \sum_{k=0}^{\infty} s_k(nx) \int_{k/n}^{(k+1)/n} f(a_k t) dt, \quad x \in [0, \infty), \tag{2.3.4}$$

where $a_k = 2k/(2k+1)$. These operators reproduce affine functions. Another modification of the Szász–Kantorovich operators was used in [14] and [47], but the operators are positive for $x \geq 0.5$.

In [31], for $b \geq 0$, the authors considered $\rho(x) = 1/(1+x)^b$ and $\varphi(x) = \sqrt{x}$. Moreover,

$$C_\rho[0, \infty) := \{f \in C[0, \infty) : \|\rho f\| < \infty\}.$$

The following direct results were discussed in [31]:

Theorem 2.6 *For a function $f \in C_\rho[0, \infty)$, one has*

$$\lim_{n \to \infty} \|\rho((\overline{K}_n^S f) - f)\| = 0$$

if and only if $(\rho f) \circ e_2$ is uniformly continuous on $[0, \infty)$.

For $0 \leq \beta \leq 1$,

$$C_{\rho,\beta}[0, \infty) := \{h \in C[0, \infty) : h(0) = 0, \|\rho \varphi^{2\beta} h\| < \infty\}.$$

Also, for $\alpha, \beta \in [0, 1]$ and $f \in C_{\rho,\beta}[0, \infty)$, the K-functional is defined as

$$K_{\alpha,\beta}(f, t)_\rho = \inf\{||\rho\varphi^{2\beta}(f - g)|| + t||\rho\varphi^{2(\alpha+\beta)}g''|| : g \in D(\alpha, \beta)\},$$

where

$$D(\alpha, \beta) = \{g \in C_{\rho,\beta}[0, \infty) : g' \in AC_{loc} : ||\rho\varphi^{2(\alpha+\beta)}g''|| < \infty\}.$$

Theorem 2.7 *If $\beta \in [0, 1]$ and $b \geq 1$, there exists a constant C such that, for all $n > 2(1 + d)$ and every $f \in C_{\rho,\beta}[0, \infty)$, we have*

$$||\varphi^{2\beta}(\overline{K}_n^S f)|| \leq C||\rho\varphi^{2\beta}f||.$$

Theorem 2.8 *If $\alpha, \beta \in [0, 1]$, $\alpha + \beta < 2$ and $b \geq 2$, then there exists a constant C such that, for all $n \geq 2(1 + d)$, every $f \in C_{\rho,\beta}[0, \infty)$ and $x > 0$*

$$\rho(x)\varphi^{2\beta}(x)|(\overline{K}_n^S f)f(x)| \leq CK_{\alpha,\beta}\left(f, \frac{\varphi^{(2(1-\alpha)}(x)}{n}\right)_\rho.$$

In another recent paper, Bustamante [32] proposed certain modified Baskakov–Kantorovich operators reproducing affine functions and established some direct results. In Baskakov operators the weights are changed while obtaining the results.

2.4 General Class of Integral Operators

For $x \in [0, \infty)$, Gupta [68] proposed the following general family of linear positive operators:

$$(V_{n,\alpha,\beta}f)(x) = n\sum_{k=1}^{\infty} m_{n,k}^\alpha(x) \int_0^\infty m_{n,k-1}^{\beta+1}(t)f(t)dt$$

$$+ m_{n,0}^\alpha(x)f(0), \tag{2.4.1}$$

where

$$m_{n,k}^\alpha(x) = \frac{(\alpha)_k \cdot \alpha^\alpha}{k!}\frac{(nx)^k}{(\alpha + nx)^{\alpha+k}}, \quad m_{n,k-1}^{\beta+1}(t) = \frac{(\beta + 1)_{k-1} \cdot \beta^{\beta+1}}{(k - 1)!}\cdot\frac{(nt)^{k-1}}{(\beta + nt)^{\beta+k}}$$

with the rising factorial $(\alpha)_k = \prod_{i=0}^{k-1}(\alpha + i)$, $(\alpha)_0 = 1$. Here, $m_{n,k}^\alpha$ is the Miheşan [121] basis function, which may also be obtained from the Mastroianni basis function [120], by substitution. We have some of the following special cases of our operators (2.4.1):

1. If $\alpha = \beta = n$ and $x \in [0, \infty)$, we obtain the Baskakov–Durrmeyer type operators (see [83]):

$$(V_{n,n,n} f)(x) = n \sum_{k=1}^{\infty} v_{n,k}(x) \int_0^{\infty} v_{n+1,k-1}(t) + (1+x)^{-n} f(0),$$

where $v_{n,k}(x)$ is the Baskakov basis given in (1.1.6).

2. If $\alpha = \beta = -n$ and $x \in [0, 1]$, we get the Bernstein–Durrmeyer polynomials (see [85]):

$$(V_{n,-n,-n} f)(x) = n \sum_{k=1}^{n} p_{n,k}(x) \int_0^1 p_{n-1,k-1}(x) f(t)dt + (1-x)^n f(0),$$

where $p_{n,k}(x)$ is the Bernstein basis given in (1.1.5).

3. If $\alpha = \beta \to \infty$ and $x \in [0, \infty)$, we get the Phillips operators (cf. [141], [119] and [50]) as follows:

$$(V_{n,\infty,\infty} f)(x) = n \sum_{k=1}^{\infty} s_k(nx) \int_0^{\infty} s_{k-1}(nt) f(t)dt + e^{-nx} f(0),$$

where $s_k(nx)$ is the Szász–Mirakyan basis function given in (1.1.7).

4. If $\alpha = \beta$ and $x \in [0, \infty)$, we get the well-known Srivastava–Gupta type operators (cf. [151] and [105]).

5. If $\alpha \neq \beta$. $\alpha = n, \beta \to \infty$ and $x \in [0, \infty)$, we get the Baskakov–Szász type operators (see [16]):

$$(V_{n,n,\infty} f)(x) = n \sum_{k=1}^{\infty} v_{n,k}(x) \int_0^{\infty} s_{k-1}(nt) f(t)dt + (1+x)^{-n} f(0),$$

where $v_{n,k}(x)$ and $s_{k-1}(nt)$ are, respectively, the Baskakov and Szász–Mirakyan basis functions defined above.

6. If $\alpha \neq \beta$ and $\alpha \to \infty, \beta = n$ and $x \in [0, \infty)$, we get the Szász–Beta type operators (see [130, (1)]):

$$(V_{n,\infty,n} f)(x) = \sum_{k=1}^{\infty} s_k(nx) \int_0^{\infty} \widetilde{b}_{n,k}(t) f(t)dt + e^{-nx} f(0),$$

where $s_k(nx)$ is defined above and the Beta basis is given by

$$\widetilde{b}_{n,k}(t) = \frac{1}{B(n,k)} \frac{t^{k-1}}{(1+t)^{n+k}}.$$

7. If $\alpha \neq \beta, \alpha = nx, \beta = n$ and $x \in [0, \infty)$, we get the Lupaş–Beta operators (see [94, (3)]):

$$(V_{n,nx,n} f)(x) = \sum_{k=1}^{\infty} l_k(nx) \int_0^{\infty} \widetilde{b}_{n,k}(t) f(t) dt + 2^{-nx} f(0),$$

where the Lupaş basis function is given by

$$l_k(nx) = 2^{-nx} \frac{(nx)_k}{k!.2^k}$$

and $\widetilde{b}_{n,k}(t)$ is the Beta basis given above.

8. If $\alpha \neq \beta$ and $\alpha = nx, \beta \to \infty$ and $x \in [0, \infty)$, we get the Lupaş–Szász type operators (see [61, (4.3)]):

$$(V_{n,nx,\infty} f)(x) = n \sum_{k=1}^{\infty} l_k(nx) \int_0^{\infty} s_{k-1}(nt) f(t) dt + 2^{-nx} f(0),$$

where $l_k(nx)$ is defined above and $s_{k-1}(nt)$ is the Szász–Mirakyan basis function.

Very recently, an asymptotic formula and the rate of convergence for functions having derivatives of bounded variation were discussed in [140]. In [75], Gupta considered a modified form of the special case of similar Bernstein–Durrmeyer type operators, which preserve linear functions and estimated approximation results in complex setting.

2.5 Operators Preserving Linear Functions

Depending on certain parameters α, β, γ and $\rho > 0$ as indicated below, Gupta [72] introduced a general family of linear positive operators, which include several operators as special cases. For $x \geq 0$, the operators are defined by

$$(A_{n,\alpha}^{\beta,\rho} f)(x) = \sum_{k=1}^{\infty} m_{n,k}^{\alpha}(x) \int_0^{\infty} m_{n,k-1}^{\beta+1,\rho}(t) f(t) dt + m_{n,0}^{\alpha}(x) f(0), \quad (2.5.1)$$

where $m_{n,k}^{\alpha}(x)$ is defined in (2.4.1) and

$$m_{n,k-1}^{\beta+1,\rho}(t) = \frac{n}{\beta.B(k\rho, \beta\rho + 1)} \cdot \frac{\left(\frac{nt}{\beta}\right)^{k\rho-1}}{\left(1 + \frac{nt}{\beta}\right)^{\beta\rho+k\rho+1}},$$

with the rising factorial $(\alpha)_k = \alpha(\alpha + 1) \cdots (\alpha + k - 1)$ where $(\alpha)_0 = 1$. These operators reproduce linear functions. Below are some special cases of the operators (2.5.1):

1. If $\alpha = \beta, \rho = 1$, we obtain the general operators, which preserve linear functions (see [73]).
2. If $\alpha = \beta \to \infty, \rho = 1$, we get the well-known Phillips operators (see [141] and [50]).
3. If $\alpha = \beta = n, \rho = 1$, we immediately get the Baskakov–Durrmeyer type operators preserving linear functions considered by Finta in [49].
4. If $\alpha \neq \beta$ and $\alpha = n, \beta \to \infty, \rho = 1$, we obtain the Baskakov–Szász type operators, which were proposed in [16].
5. If $\alpha \neq \beta$ and $\alpha \to \infty, \beta = n, \rho = 1$, we get the Szász–Beta type operators proposed in [86].
6. If $\alpha \neq \beta$ and $\alpha = nx, \beta = n, \rho = 1$, we get the Lupaş–Beta operators introduced in [94, (1.2)].
7. If $\alpha \neq \beta$ and $\alpha = nx, \beta \to \infty, \rho = 1$, we get immediately the Lupaş–Szász type operators proposed in [61, (4.3)].
8. If $\alpha = \beta = -n, \rho = 1$, we obtain the genuine Bernstein–Durrmeyer operators, introduced by Chen [33] as well as by Goodman–Sharma [60].
9. If $\alpha = \beta = n, \rho > 0$, we get the link operators due to Heilmann and Rasa [101, (2)] for $c = 1$.
10. If $\alpha = \beta \to \infty, \rho > 0$, we get the Păltănea operators [138]. Clearly,

$$m_{n,k}^{\infty}(x) = e^{-nx} \frac{(nx)^k}{k!},$$

and for the weights we have

$$\lim_{\beta \to \infty} m_{n,k-1}^{\beta+1,\rho}(t) = \frac{(n\rho)^{k\rho} t^{k\rho-1} e^{-n\rho t}}{\Gamma(k\rho)},$$

which is the form recently considered by M. Heilmann and I Raşa [101], but for our operators we observe that one has to consider $k\rho$ as a positive integer, which is the class of operators for the parameter ρ bigger than those for $\rho \in \mathbb{N}$.

11. If $\beta \to \infty, \rho > 0$ and for each α, with $k\rho$ being a positive integer, we get the Miheşan–Durrmeyer type operators proposed by Kajla in [109].
12. If $\alpha = \beta = -n, \rho > 0$, we obtain the link Bernstein–Durrmeyer operators preserving linear functions, introduced in [139] and further studied in [57]. For this special case by simple computations, we obtain

$$m_{n,k}^{-n}(x) = \binom{n}{k} x^k (1 - x)^{n-k},$$

and for the weights we proceed as follows:

$$m_{n,k-1}^{\beta+1,\rho}(t) = \frac{n(\beta\rho + k\rho)(\beta\rho + k\rho - 1)\cdots(\beta\rho + 1)}{\beta\Gamma(k\rho)} \cdot \frac{\left(\frac{nt}{\beta}\right)^{k\rho-1}}{\left(1 + \frac{nt}{\beta}\right)^{\beta\rho+k\rho+1}}.$$

By the substitution of $\beta = -n$ for $1 \le k \le n - 1$ and considering $\mathring{k}\rho$ as a positive integer, we immediately get from our general operators the genuine link Bernstein–Durrmeyer operators

$$m_{n,k-1}^{-n+1,\rho}(t) = \frac{1}{B(k\rho, (n-k)\rho)} \cdot t^{k\rho-1}(1-t)^{n\rho-k\rho-1}.$$

Thus, in order to preserve the constant function, the operators for $x \in [0, 1]$ take the following form of the general operators (2.5.1):

$$(A_{n,\alpha}^{\beta,\rho}f)(x) \equiv (A_{n,-n}^{-n,\rho}f)(x) = \sum_{k=0}^{n} m_{n,k}^{-n}(x)G_{n,k}^{-n,\rho}(f), \qquad (2.5.2)$$

with

$$G_{n,k}^{-n,\rho}(f) = \begin{cases} \int_0^1 m_{n,k-1}^{-n+1,\rho}(t)f(t)dt, & 1 \le k \le n-1 \\ f(0), & k = 0 \\ f(1), & k = n. \end{cases}$$

13. Corresponding to the cases (4)–(7) mentioned above, we may define the hybrid link operators for $\rho > 0$ and consider the different values of α and β as in these cases.

Remark 2.4 We point out here that for some more special cases, when $\beta = nt$, one can consider many operators which provide the weights of Lupaş basis functions. But, such operators are not appropriate from a convergence point of view.

We consider here another new hybrid operator, which for $x \in [0, \infty)$ is defined as

$$(\tilde{V}_n f)(x) = (n+1)\sum_{k=1}^{\infty} r_{n,k}(x)\int_0^{\infty} v_{n+2,k-1}(t)f(t)dt + r_{n,0}(x)f(0),$$

where $r_{n,k}(x)$ and $v_{n,k}(t)$ are defined in (1.1.11) and (2.7.1). These operators preserve constant and linear functions, which can be seen from the following lemmas.

Lemma 2.1 *If $U_{n,m}(x) = \sum_{k=0}^{\infty} r_{n,k}(x) \left(\frac{k}{n} - x\right)^m$, $m \in N \cup \{0\}$, then the moments satisfy the following recurrence relation:*

$$nU_{n,m+1}(x) = x(1+x)^2 \left[(U_{n,m}(x))' + mU_{n,m-1}(x)\right].$$

In general, we have $U_{n,m}(x) = O_x(n^{-[(m+1)/2]})$, where $[s]$ stands for the integral part of s.

Lemma 2.2 ([90]) *If we denote $\mu_{n,m}(x) = (\widetilde{V}_n(e_1 - xe_0)^m)(x)$ for $m \in N$ and $x \geq 0$, then*

$$(n - m)\mu_{n,m+1}(x) = x(1+x)^2 \left[[\mu_{n,m}(x)]' + m\mu_{n,m-1}(x)\right]$$
$$= +m\mu_{n,m}(x) + mx(1+x)\mu_{n,m-1}(x).$$

In particular, we have

$$\mu_{n,0}(x) = 1, \ \mu_{n,1}(x) = 0, \ \mu_{n,2}(x) = \frac{x(1+x)(2+x)}{n-1}.$$

In general, for all $x \in [0, \infty)$, we have $\mu_{n,m}(x) = O_x(n^{-[(m+1)/2]})$.

Lemma 2.3 ([90]) *If we denote $T_{n,m}(x) = (\widetilde{V}_n e_m)(x)$ for $m \in N$ and $x \geq 0$, then*

$$(n - m)T_{n,m+1}(x) = x(1+x)^2[T_{n,m}(x)]' + (m + nx)T_{n,m}(x).$$

In particular, we have

$$T_{n,0}(x) = 1, \ T_{n,1}(x) = x, \ T_{n,2}(x) = x^2 + \frac{x(1+x)(2+x)}{n-1}.$$

$$T_{n,3}(x) = x^3 + \frac{3x^2(1+x)(2+x)}{n-2} + O_x(n^{-2}).$$

In general, we have

$$T_{n,r}(x) = x^r + \frac{r(r-1)}{2} x^{r-1} \frac{(1+x)(2+x)}{n-r+1} + O_x(n^{-2}).$$

Lemma 2.4 ([79]) *There exist polynomials $q_{i,j,r}(x)$ independent of n and k such that*

$$[x^r(1+x)^{2r}]\frac{\partial^r}{\partial x^r}[r_{n,k}(x)] = \sum_{\substack{2i+j \leq r \\ i,j \geq 0}} n^i(k - nx)^j q_{i,j,r}(x)[r_{n,k}(x)].$$

Using the above lemmas and following the methods given in [79], we have the asymptotic formula in simultaneous approximation as indicated below:

Theorem 2.9 ([90]) *Let* $f \in C[0, \infty)$ *with* $|f(t)| \leq C(1 + t)^\gamma$ *for some* $\gamma > 0$, $t \geq 0$. *If* $f^{(r+2)}$ *exists at a point* $x \in (0, \infty)$, *then we have*

$$
\lim_{n \to \infty} n((\widetilde{V}_n^{(r)} f)(x) - f^{(r)}(x)) = \frac{r(r-1)(r-2)}{2} f^{(r-1)}(x)
$$
$$
+ \frac{3r(r-1)(x+1)}{2} f^{(r)}(x)
$$
$$
+ \frac{r[(r^2+2)x^2 + 3(r+1)x + 2]}{2} f^{(r+1)}(x)
$$
$$
+ \frac{x(1+x)(2+x)}{2} f^{(r+2)}(x).
$$

The proof of this theorem follows using Lemmas 2.1, 2.2, 2.3 and 2.4 (also see [79, Th. 3]).

2.6 Usual Durrmeyer Type and Mixed Hybrid Operators

For $x \in \mathbb{R}^+ \equiv [0, \infty)$ and α and β non-zero real numbers or a function of n (as indicated below), Gupta in [64] introduced the usual Durrmeyer type operators as

$$
(D_{n,\alpha,\beta} f)(x) = \int_0^\infty \phi_{n,\alpha,\beta}^D(x, t) f(t) dt
$$
$$
= \sum_{k=0}^\infty m_{n,k}^\alpha(x) G_{n,k}^\beta(f), \tag{2.6.1}
$$

where the kernel is

$$
\phi_{n,\alpha,\beta}^D(x, t) = \sum_{k=0}^\infty m_{n,k}^\alpha(x) m_{n,k}^\beta(t),
$$

$G_{n,k}^\beta(f) = \frac{n(\beta-1)}{\beta} \langle m_{n,k}^\beta, f \rangle$ with $\langle f, g \rangle = \int_0^\infty f(t) g(t) dt$ and $m_{n,k}^\alpha(x)$ defined in (2.4.1).

1. If $\alpha = \beta = n/c$, $c \in \mathbb{N}_0$, we get the well-known operators due to Heilmann–Müller (see [100]).
2. If $\alpha = \beta = n$, we get the Baskakov–Durrmeyer operators considered in [147].
3. If $\alpha = \beta \to \infty$, we get the Szász–Durrmeyer operators (see [123]).

4. If $\alpha = \beta = -n$, we get the Bernstein–Durrmeyer polynomials introduced in [48], in this case $x \in [0, 1]$ and the summation is for $0 \le k \le n$.
5. If $\alpha \ne \beta$ and $\alpha = n, \beta \to \infty$, we get the Baskakov–Szász operators [93].
6. If $\alpha \ne \beta$ and $\alpha \to \infty, \beta = n$, we get the Szász–Baskakov operators [142].
7. If $\alpha \ne \beta$ and $\alpha = nx, \beta = n$, we get the Lupaş–Baskakov operators [61] for the case $d = 1, c = 1$.
8. If $\alpha \ne \beta$ and $\alpha = nx, \beta \to \infty$, we get the Lupaş–Szász operators [61] for the case $d = 0, c = 1$.
9. If $\alpha \ne \beta$ and $\alpha = nx, \beta = nt$, we get the Lupaş–Durrmeyer operators considered by Agratini [13], but in this case $G_{n,k}^{\beta}(f) \ne \frac{n(\beta-1)}{\beta} \langle m_{n,k}^{\beta}, f \rangle$ as $\langle m_{n,k}^{\beta}, 1 \rangle \ne \frac{n(\beta-1)}{\beta}$. Thus, from a convergence point of view, this case is not suitable.

Following [64], we have

$$\phi_{n,\alpha,\beta}^D(x, t) = \left(1 + \frac{nx}{\alpha}\right)^{-\alpha} \left(1 + \frac{nt}{\beta}\right)^{-\beta} {}_2F_1\left(\alpha, \beta; 1; \frac{n^2xt}{(\alpha + nx)(\beta + nt)}\right),$$

where

$$ {}_2F_1(a, b; c; x) = \sum_{r=0}^{\infty} \frac{(a)_r(b)_r}{(c)_r} \frac{x^r}{r!}, \quad |x| < 1.$$

This kernel can be linked to some special functions for different values of α and β. Some of these links can be observed in various places separately (see, for example, [27]):

1. If $\alpha = \beta \to \infty$, we get the kernel of the Szász–Durrmeyer operators

$$\lim_{\alpha,\beta\to\infty} \phi_{n,\alpha,\beta}^D(x, t) = e^{-n(x+t)} \sum_{v=0}^{\infty} \frac{(n^2xt)^v}{v!\Gamma(v + 1)} = e^{-n(x+t)} I_0(2n\sqrt{xt}),$$

where I_0 is the modified Bessel's function of the first kind, of zero order.
2. If $\alpha = n, \beta \to \infty$, we get the kernel of the Baskakov–Szász operators

$$\lim_{\alpha\to n,\beta\to\infty} \phi_{n,\alpha,\beta}^D(x, t) = (1 + x)^{-n} e^{-nt} e^{\frac{nxt}{1+x}} {}_1F_1\left(1 - n; 1; \frac{-nxt}{1 + x}\right)$$

$$= (1 + x)^{-n} e^{\frac{-nt}{1+x}} L_{n-1}^0\left(\frac{-nxt}{1 + x}\right),$$

where ${}_1F_1(a; b; x) = \sum_{r=0}^{\infty} \frac{(a)_r}{(b)_r} \frac{x^r}{r!}$ is the confluent hypergeometric function.
3. If $\alpha \to \infty, \beta = n$, we get the kernel of the Szász–Baskakov operators as follows:

$$\lim_{\alpha\to\infty,\beta\to n} \phi_{n,\alpha,\beta}^D(x, t) = (1 + t)^{-n} e^{\frac{-nx}{1+t}} L_{n-1}^0\left(\frac{-nxt}{1 + t}\right).$$

4. If $\alpha = nx, \beta \to \infty$, we get the kernel of the Lupaş–Szász operators:

$$\lim_{\alpha \to nx, \beta \to \infty} \phi_{n,\alpha,\beta}^D(x,t) = 2^{-nx} e^{-nt/2} L_{nx-1}^0 \left(\frac{-nt}{2} \right),$$

where $L_m^0(x)$ in the above cases are generalized Laguerre functions.

Lemma 2.5 *The r-th ($r \in \mathbb{N}$) order moment with $e_r(t) = t^r$ (for the cases (1)–(8) of (2.6.1)) can be represented as*

$$(D_{n,\alpha,\beta} e_r)(x) = \frac{\Gamma(\beta - r - 1)\Gamma(r+1)}{\Gamma(\beta - 1)} \left(\frac{\beta}{n} \right)^r {}_2F_1 \left(\alpha, -r; 1; \frac{-nx}{\alpha} \right).$$

In particular,

$$(D_{n,\alpha,\beta} e_0)(x) = 1, \quad (D_{n,\alpha,\beta} e_1)(x) = \frac{\beta(1+nx)}{n(\beta-2)},$$

$$(D_{n,\alpha,\beta} e_2)(x) = \frac{\beta^2[2\alpha + 4\alpha nx + (\alpha+1)n^2x^2]}{\alpha n^2(\beta-2)(\beta-3)}.$$

2.7 Some New Operators

For $\alpha = \alpha(n,x), \beta = \beta(n,x)$ and $\eta = \eta(n)$ under the conditions $0 < \alpha < \infty, |\beta| < 1$, we propose the following representation for $x \in [0, \infty)$ having hybrid form of basis functions as

$$(G_n^{\alpha,\beta,\eta} f) = \frac{n(\eta-1)}{\eta} \sum_{k=0}^{\infty} \omega_\beta(k,\alpha) \int_0^\infty m_{n,k}^\eta(x) f(t) dt, \qquad (2.7.1)$$

where

$$\omega_\beta(k,\alpha) = \frac{\alpha(\alpha+k\beta)^{k-1}}{k!} e^{-(\alpha+k\beta)}, m_{n,k}^\eta(x) = \frac{(\eta)_k \cdot \eta^\eta}{k!} \frac{(nx)^k}{(\eta+nx)^{\eta+k}}.$$

Also, we use the notations

$$r_{n,k}(x) = e^{-(n+k)x/(1+x)} \frac{n(n+k)^{k-1}}{k!} \left(\frac{x}{1+x} \right)^k,$$

$$j_{n,k}^\beta(x) = \frac{nx(nx+k\beta)^{k-1} e^{-(nx+k\beta)}}{k!},$$

$$s_k(nx) = \frac{e^{-nx}(nx)^k}{k!},$$

$$v_{n,k}(x) = \binom{n+k-1}{k}\frac{t^k}{(1+t)^{n+k}}.$$

Some of the particular cases of the representation (2.7.1) provide many well-known operators as special cases, some of which are listed below:

1. If $\alpha = nx, \eta \to \infty$, we get the hybrid operators having the Jain–Szász basis functions as

$$(G_n^{nx,\beta,\infty}f)(x) = n\sum_{k=0}^{\infty}j_{n,k}^{\beta}(x)\int_0^{\infty}s_k(nx)f(t)dt.$$

2. If $\alpha = nx, \beta = 0, \eta \to \infty$, we get the well-known Szász–Durrmeyer operators

$$(G_n^{nx,0,\infty}f)(x) = n\sum_{k=0}^{\infty}s_{n,k}(x)\int_0^{\infty}s_k(nt)f(t)dt.$$

3. If $\alpha = nx, \eta = n$, we immediately get the operators having the Jain–Baskakov basis as

$$(G_n^{nx,\beta,n}f)(x) = (n-1)\sum_{k=0}^{\infty}j_{n,k}^{\beta}(x)\int_0^{\infty}v_{n,k}(t)f(t)dt.$$

4. If $\alpha = nx, \beta = 0, \eta = n$, we may obtain the well-known Szász–Baskakov operators:

$$(G_n^{nx,0,n}f)(x) = (n-1)\sum_{k=0}^{\infty}s_{n,k}(x)\int_0^{\infty}v_{n,k}(t)f(t)dt.$$

5. If $\alpha = nx/(1+x), \beta = x/(1+x), \eta \to \infty$, we get a new operator having the Ismail–May–Szász basis functions

$$(G_n^{nx/(1+x),x/(1+x),\infty}f)(x) = n\sum_{k=0}^{\infty}r_{n,k}(x)\int_0^{\infty}s_k(nt)f(t)dt.$$

6. If $\alpha = nx/(1+x), \beta = x/(1+x), \eta = n$, we derive another new operator with the Ismail–May–Baskakov basis functions

$$(G_n^{nx/(1+x),x/(1+x),n}f)(x) = (n-1)\sum_{k=0}^{\infty}r_{n,k}(x)\int_0^{\infty}v_{n,k}(x)f(t)dt.$$

In order to obtain the moments of these operators, we use the following two lemmas followed by the remark.

Lemma 2.6 ([107]) *If*

$$S(r, \alpha, \beta) = \sum_{k=0}^{\infty} \frac{(\alpha + k\beta)^{k+r-1}}{k!} e^{-(\alpha+k\beta)},$$

then $\alpha S(0, \alpha, \beta) = 1$, *and*

$$S(r, \alpha, \beta) = \sum_{k=0}^{\infty} \beta^k (\alpha + k\beta) S(r - 1, \alpha + k\beta, \beta).$$

Additionally, we have the recurrence relation:

$$S(r, \alpha, \beta) = \alpha S(r - 1, \alpha, \beta) + \beta S(r, \alpha + \beta, \beta).$$

Lemma 2.7 *For the representation (1.2.1), we have*

$$(S_{\alpha,\beta}^n e_0)(x) = 1$$

$$(S_{\alpha,\beta}^n e_1)(x) = \frac{\alpha}{n(1 - \beta)}$$

$$(S_{\alpha,\beta}^n e_2)(x) = \frac{\alpha}{n^2} \left[\frac{\alpha}{(1 - \beta)^2} + \frac{1}{(1 - \beta)^3} \right]$$

$$(S_{\alpha,\beta}^n e_3)(x) = \frac{\alpha}{n^3} \left[\frac{\alpha^2}{(1 - \beta)^3} + \frac{3\alpha}{(1 - \beta)^4} + \frac{(1 + 2\beta)}{(1 - \beta)^5} \right]$$

$$(S_{\alpha,\beta}^n e_4)(x) = \frac{\alpha}{n^4} \left[\frac{\alpha^3}{(1 - \beta)^4} + \frac{6\alpha^2}{(1 - \beta)^5} + \frac{(7 + 8\beta)\alpha}{(1 - \beta)^6} + \frac{(1 + 8\beta + 6\beta^2)}{(1 - \beta)^7} \right].$$

The proof of Lemma 2.7 follows using Lemma 2.6 and the methods used in [107].

Remark 2.5 For the representation (2.7.1), using the fact $(\eta)_k = \frac{\Gamma(\eta+k)}{\Gamma(\eta)}$, we obtain that

$$\frac{n(\eta - 1)}{\eta} \int_0^\infty m_{n,k}^\eta(x) e_r(t) dt = \left(\frac{\eta}{n} \right)^r \frac{\Gamma(\eta - r - 1)}{\Gamma(\eta - 1)} \frac{(k + r)!}{k!}.$$

Lemma 2.8 *From (2.7.1), Remark 2.5 and Lemma 2.7, we have the following:*

$$(G_n^{\alpha,\beta,\eta} e_0)(x) = 1$$

$$(G_n^{\alpha,\beta,\eta} e_1)(x) = \frac{\eta}{\eta - 2} \left[\frac{\alpha}{n(1 - \beta)} + \frac{1}{n} \right]$$

$$G_n^{\alpha,\beta,\eta}(e_2, x) = \frac{\eta^2}{(\eta-2)(\eta-3)}\left[\frac{\alpha^2}{n^2(1-\beta)^2} + \frac{\alpha}{n^2(1-\beta)^3} + \frac{3\alpha}{n^2(1-\beta)} + \frac{2}{n^2}\right]$$

$$G_n^{\alpha,\beta,\eta}(e_3, x) = \frac{\eta^3}{(\eta-2)(\eta-3)(\eta-4)}\left[\frac{\alpha^3}{n^3(1-\beta)^3} + \frac{3\alpha^2}{n^3(1-\beta)^4} + \frac{\alpha(1+2\beta)}{n^3(1-\beta)^5}\right.$$
$$\left. + \frac{6\alpha^2}{n^3(1-\beta)^2} + \frac{6\alpha}{n^3(1-\beta)^3} + \frac{11\alpha}{n^3(1-\beta)} + \frac{6}{n^3}\right]$$

$$G_n^{\alpha,\beta,\eta}(e_4, x) = \frac{\eta^4}{(\eta-2)(\eta-3)(\eta-4)(\eta-5)}\left[\frac{\alpha^4}{n^4(1-\beta)^4} + \frac{6\alpha^3}{n^4(1-\beta)^5}\right.$$
$$+ \frac{(7+8\beta)\alpha^2}{n^4(1-\beta)^6} + \frac{(1+8\beta+6\beta^2)\alpha}{n^4(1-\beta)^7}$$
$$+ \frac{10\alpha^3}{n^4(1-\beta)^3} + \frac{30\alpha^2}{n^4(1-\beta)^4} + \frac{10\alpha(1+2\beta)}{n^4(1-\beta)^5}$$
$$\left. + \frac{35\alpha^2}{n^4(1-\beta)^2} + \frac{35\alpha}{n^4(1-\beta)^3} + \frac{50\alpha}{n^4(1-\beta)} + \frac{24}{n^4}\right].$$

Using Lemma 2.8, one can find the moments for the different cases as indicated above in (2.7.1) by assigning different values to α and η. It is observed that all the six cases indicated above preserve only the constant functions.

In [84], Gupta and Greubel defined the usual Durrmeyer variant of the similar operators as follows:

$$(D_n^\beta f)(x) = \sum_{k=0}^{\infty} \left(\int_0^{\infty} (j_{n,k}^\beta(t)dt)^{-1} j_{n,k}^\beta(x) \int_0^{\infty} j_{n,k}^\beta(t)f(t)dt.\right.$$

We cannot replace $j_{n,k}^\beta(t)$ above with $r_{n,k}(t)$, as for such a case the operators will not converge. After considering suitable weights, one can consider the Durrmeyer variant.

Starting with the operators of Baskakov type, which for $c \geq 0$ are defined by

$$(L_{n,\infty}^{[c]} f)(x) \sum_{v=0}^{\infty} p_{n,v}^{[c]}(x) f\left(\frac{v}{n}\right),$$

where

$$p_{n,v}^{[c]}(x) = \begin{cases} \frac{(nx)^v}{v!} e^{-nx}, & \text{if } c = 0 \\ \frac{\Gamma(\frac{n}{c}+v)}{\Gamma(\frac{n}{c})\Gamma(v+1)}(cx)^v(1+cx)^{\frac{-n}{c}-v} & \text{if } c > 0. \end{cases}$$

Abel et al. [5] considered linking operators depending on a parameter $\rho > 0$ as follows:

$$(L_{n,\rho}^{[c]}f)(x) = \sum_{v=1}^{\infty} p_{n,v}^{[c]}(x) \int_0^{\infty} \mu_{n,v,\rho}^{[c]}(t)f(t)dt$$

$$+ p_{n,0}^{[c]}(x)f(0), \qquad\qquad (2.7.2)$$

where

$$\mu_{n,v,\rho}^{[c]}(t) = \frac{c}{B\left(\frac{n\rho}{c}+1, v\rho\right)}(ct)^{v\rho-1}(1+ct)^{\frac{-n\rho}{c}-v\rho-1}.$$

These operators are similar to the operators (2.5.1) and have a link with the Baskakov operators and the genuine Baskakov operators [49]. In case $c = 0$, these operators reduce to the Phillips type link operators [138]. The case $\rho = 1$ was given in [73, Ex. 2].

Abel et al. [5] obtained the following main result.

Theorem 2.10 ([5]) *Let $c, \gamma > 0$. Assume that $f \in C[0, \infty) \to R$ satisfies the growth condition $f(t) = O(t^{t\gamma})$ as $t \to \infty$. Then, for any $b > 0$, there is a constant $\rho_0 > 0$ such that $(L_{n,\rho}^{[c]}f)$ exists for all $\rho \geq \rho_0$ and*

$$\lim_{\rho\to\infty}(L_{n,\rho}^{[c]}f)(x) = (L_{n,\infty}^{[c]})(x)$$

uniformly for $x \in [0, b]$.

Additionally, in [40], the Durrmeyer variant of the Apostol–Genocchi–Baskakov type hybrid operators was considered and their approximation behaviour has been discussed.

2.8 Further Modifications

Bascanbaz-Tunca et al. [154] proposed another unified approach to define a discrete sequence of linear positive operators, which for $x \geq 0$ and $0 \leq \beta < 1$ is defined as

$$M_n^{\beta}(f, x) = \sum_{i=0}^{\infty} \frac{nx(nx+1+i\beta)_{i-1}}{2^i \cdot i!} \cdot 2^{-(nx+i\beta)} f\left(\frac{i}{n}\right). \qquad (2.8.1)$$

In [154], they termed these operators as the Jain–Lupaş operators. As a special case, if $\beta = 0$, we immediately get the operators of Lupaş, studied by Agratini [13] and defined as

$$M_n^0(f, x) = \sum_{i=0}^{\infty} \frac{(nx)_i}{2^i \cdot i!} \cdot 2^{-nx} f\left(\frac{i}{n}\right).$$

Very recently, Gupta et al. [88] introduced the following representation by considering the weight function of Miheşan (see [121]) under integration. For $x \geq 0$, $0 \leq \beta < 1$ and $\eta \in \{n\} \cup \{\infty\}$, $n \in N$, they considered

$$G_n^{\beta,\eta}(f, x) = \frac{n(\eta - 1)}{\eta} \sum_{i=0}^{\infty} \omega_\beta(i, nx) \int_0^{\infty} m_{n,i}^\eta(t) f(t) dt, x \in [0, \infty), \quad (2.8.2)$$

where

$$\omega_\beta(i, nx) = \frac{nx(nx + 1 + i\beta)_{i-1}}{2^i \cdot i!} \cdot 2^{-(nx+i\beta)}, \quad m_{n,i}^\eta(t) = \frac{(\eta)_i}{i!} \cdot \frac{\left(\frac{nt}{\eta}\right)^i}{\left(1 + \frac{nt}{\eta}\right)^{n+i}},$$

which provide many operators of summation-integral type with different basis. Some of the special cases of (2.8.2) are indicated in the following:

1. If $\eta \to \infty$, we derive the Jain–Lupaş–Szász operators, which for $x \in [0, \infty)$ are defined as

$$G_n^{\beta,\infty}(f, x) = n \sum_{i=0}^{\infty} \frac{nx(nx + 1 + i\beta)_{i-1}}{2^i \cdot i!} \cdot 2^{-(nx+i\beta)} \int_0^{\infty} e^{-nt} \frac{(nt)^i}{i!} f(t) dt.$$

2. If $\beta = 0, \eta \to \infty$, we obtain the Lupaş–Szász operators, for $x \in [0, \infty)$ defined by

$$G_n^{0,\infty}(f, x) = n \sum_{i=0}^{\infty} \frac{(nx)_i}{2^i \cdot i!} \cdot 2^{-nx} \int_0^{\infty} e^{-nt} \frac{(nt)^i}{i!} f(t) dt.$$

3. If $\eta = n$, we immediately get the Jain–Lupaş–Baskakov operators, defined in $x \in [0, \infty)$ by

$$G_n^{\beta,n}(f, x) = (n - 1) \sum_{i=0}^{\infty} \frac{nx(nx + 1 + i\beta)_{i-1}}{2^i \cdot i!} \cdot 2^{-(nx+i\beta)}$$

$$\int_0^{\infty} \binom{n + i - 1}{i} \frac{t^i}{(1 + t)^{n+i}} f(t) dt.$$

4. If $\beta = 0, \eta = n$, we derive the Lupaş–Baskakov operators, defined for $x \in [0, \infty)$ as

$$G_n^{0,n}(f, x) = (n-1) \sum_{i=0}^{\infty} \frac{(nx)_i}{2^i \cdot i!} \cdot 2^{-nx} \int_0^{\infty} \binom{n+i-1}{i} \frac{t^i}{(1+t)^{n+i}} f(t) dt.$$

The authors provided the following direct estimates in [88].

Theorem 2.11 ([88]) *Let f be a bounded integrable function on $[0, \infty)$ and f'' exist at a point $x \in [0, \infty)$. If $\beta = \beta(n) \to 0$, as $n \to \infty$ and $\lim_{n\to\infty} n\beta(n) = l \in R$, then*

$$\lim_{n\to\infty} n \left[G_n^{\beta,\eta}(f, x) - f(x) \right] = (1 + lx) f'(x) + \frac{x(x+3)}{2} f''(x).$$

Theorem 2.12 ([88]) *For $f \in C_B[0, \infty)$ and for all $x \in [0, \infty)$, there exists a constant $C > 0$, such that*

$$|G_n^{\beta,\eta}(f, x) - f(x)| \le C\omega_2(f, \sqrt{h_m}) + \omega\left(f, \left| \frac{\eta}{(\eta-2)} \left[\frac{x}{1-\beta} + \frac{1}{n} \right] - x \right| \right),$$

where

$$h_m = \mu_{n,2}^{\beta,\eta}(x) + \left(\frac{\eta}{(\eta-2)} \left[\frac{x}{1-\beta} + \frac{1}{n} \right] - x \right)^2.$$

For the weight function $\varphi(x) = 1 + x^2$, we consider the space

$$B_\varphi([0, \infty)) = \{ f : [0, \infty) \to \mathbf{R} : \quad |f(x)| \le M_f \varphi(x), x \in [0, \infty) \},$$

where M_f is a positive constant depending only on f. The space $B_\varphi([0, \infty))$ is endowed with the norm

$$\| f(x) \|_\varphi = \sup_{0 \le x < \infty} \frac{|f(x)|}{\varphi(x)}.$$

We consider the notations

$$C_\varphi([0, \infty)) = \{ f \in B_\varphi([0, \infty)) : f \text{ is continuous} \}$$

and

$$C_\varphi^k([0, \infty)) = \left\{ f \in C_\varphi([0, \infty)) : \lim_{x\to\infty} \frac{|f(x)|}{\phi(x)} = k_f \right\},$$

where k_f is also a positive constant depending only on f.

The following theorem is on weighted approximation guided by Gadjiev's Korovkin type theorem (see [51]).

Theorem 2.13 ([88]) *For $\{G_n^{\beta,\eta}\}$ being a sequence of linear and positive operators and for each $f \in C_\varphi^k([0,\infty))$, we have*

$$\lim_{n\to\infty} \|G_n^{\beta,\eta}(f,x) - f(x)\|_\varphi = 0.$$

In order to establish an asymptotic expansion, Abel–Ivan [7] considered the Pethe and Jain [108] operators in a slightly different form, as follows:

$$M_{n,c}^0(f,x) = \sum_{i=0}^{\infty} \left(\frac{c}{1+c}\right)^i \frac{(ncx)_i}{(1+c)^{ncx} \cdot i!} f\left(\frac{i}{n}\right), \qquad (2.8.3)$$

satisfying $c = c_n > \alpha$ for a certain constant $\alpha > 0$. It was observed in [7] that the operators $M_{n,c}^0$ are well defined, for all sufficiently large n, since the infinite sum in (2.8.3) is convergent if $n > A/\log(1+c)$, provided that $|f(t)| \le Ke^{At}, t \ge 0$. Gupta et al. [88] established a further generalization of these operators in the sense of Bascanbaz-Tunca et al. [154], as follows:

$$M_{n,c}^\beta(f,x) = \sum_{i=0}^{\infty} \frac{ncx(ncx+1+i\beta)_{i-1}}{(1+c)^{ncx+i\beta} \cdot i!} \left(\frac{c}{1+c}\right)^i f\left(\frac{i}{n}\right). \qquad (2.8.4)$$

If $\beta = 0$, we get the operators (2.8.3) and if $c = 1$, we obtain the generalization due to [154].

In terms of weighted modulus of continuity (see Ispir and Atakut [25]), the following result is also formulated.

Theorem 2.14 ([88]) *If*

$$f, \; f'' \in C_2^*[0,\infty) =: \left\{ f \in B_2[0,\infty) \cap C[0,\infty) : \lim_{x\to\infty} \frac{f(x)}{1+x^2} = l^* \right\}$$

with

$$B_2[0,\infty) = \{f \in [0,\infty) : |f(x)| \le M(1+x^2), M > 0\}$$

and $\beta = \beta(n) \to 0$ as $n \to \infty$, then we have for $x \in [0,\infty)$ that

$$\left| M_{n,c}^\beta(f,x) - f(x) - \left[\frac{\beta}{1-\beta} + \frac{c-1}{(c+1)(1-\beta)}\right] x f'(x) \right.$$

$$\left. - \left(\left[\frac{c(4c^3 - 3c - 2) + 1}{(1+c)^2(1-\beta)^2} + \frac{2\beta(c-1)}{(1+c)(1-\beta)^2} + \frac{\beta^2}{(1-\beta)^2}\right] x^2\right.\right.$$

$$+\frac{2xc^2\left[(\beta-1)^2+c(3+2\beta-\beta^2)\right]}{n(1+c)^2(1-\beta)^3}\bigg)f''(x)\bigg|$$

$$=8\left(1+x^2\right)O\left(n^{-1}\right)\Omega\left(f'',\frac{1}{\sqrt{n}}\right),$$

where $\Omega(f;\delta)=\displaystyle\sup_{x\in[0,\infty),|h|\le\delta}\frac{|f(x+h)-f(x)|}{(1+h^2)(1+x^2)}.$

Recently, a Gupta type variant of the Shepard operators was introduced and studied in [20] based on the Bézier variants, where the convergence results, pointwise, uniform direct and converse approximation results are given. An application to image compression improving the previous algorithm was also discussed in [20]. These investigations are extended further in [21].

Chapter 3
Difference Between Operators

3.1 Introduction

The interest towards the study of differences of positive linear operators began with the question raised by A. Lupaş [114] with regard to the possibility to establish the following estimate:

$$[B_n, \overline{\mathbb{B}}_n] := B_n \circ \overline{\mathbb{B}}_n - \overline{\mathbb{B}}_n \circ B_n = \widehat{U}_n - \widehat{S}_n,$$

where B_n are the Bernstein operators and $\overline{\mathbb{B}}_n$ are the Beta operators (see [115]).

Using Taylor's expansion with Peano remainder, Gonska et al. [58] established more general results with regard to the problem posed by Lupaş. To do so, they considered ω_k to be the k-th order modulus of smoothness, and $\widetilde{\omega}$ to be the least concave majorant of ω_1. Gonska et al. [58] estimated the following general results for difference of operators:

Theorem 3.1 Let $A, B : C[0, 1] \to C[0, 1]$ be two positive operators, such that for $x \in [0, 1]$ and for $i = 0, 1, 2, 3$ if

$$(A - B)((e_1 - xe_0)^i, x) = 0,$$

then for $f \in C^3[0, 1]$, we have

$$((A - B)f)(x) = \frac{1}{6}((A + B)|e_1 - xe_0|^3)(x)\widetilde{\omega}$$
$$\times \left(f''', \frac{1}{4} \frac{((A + B)(e_1 - xe_0)^4)(x)}{((A + B)|e_1 - xe_0|^3)(x)} \right).$$

V. Gupta, M. T. Rassias, *Computation and Approximation*, SpringerBriefs in Mathematics, https://doi.org/10.1007/978-3-030-85563-5_3

Theorem 3.2 *If A and B are given as in Theorem 3.1, also satisfying $Ae_0 = Be_0 = e_0$, then for all $f \in C[0, 1], x \in [0, 1]$ we have*

$$|((A - B)f)(x)| \le c\omega_4 \left(f, \left(\frac{1}{2}((A + B)(e_1 - xe_0)^4)(x) \right)^{1/4} \right),$$

where c is an absolute constant independent of f, x, A and B.

Based on these two theorems, Gonska [58] provided the answer of Lupaş problem:

Theorem 3.3 *If $\widehat{S}_n$ and $\widehat{U}_n$ are given as above, then*

$$|((\widehat{S}_n - \widehat{U}_n)f)(x) \le c_1\omega_4 \left(f, \left(\frac{3x(1 - x)}{n(n + 1)} \right)^{1/4} \right),$$

where c_1 is a constant independent of n, f and x.

Recently Acu–Raşa [10] estimated some interesting results for the difference of operators in order to generalize the problem of Lupaş [114] on polynomial differences. Essential results on this topic are collected in [11] regarding the difference of operators. Aral et al. [23] considered the approximation of differences of operators having the same basis functions in terms of weighted modulus of continuity.

Let us consider $F_{n,k}, G_{n,k}, H_{n,k} : D \to \mathbb{R}$, where D is a subspace of $C[0, \infty)$, which contains polynomials of degree up to 4. We define the operators

$$(U_n f)(x) = \sum_{k=0}^{\infty} v_{n,k}(x) F_{n,k}(f),$$

$$(V_n f)(x) = \sum_{k=0}^{\infty} v_{n,k}(x) G_{n,k}(f)$$

$$(W_n f)(x) = \sum_{k=0}^{\infty} u_{n,k}(x) H_{n,k}(f)$$

with $F_{n,k}(e_0) = G_{n,k}(e_0) = H_{n,k}(e_0) = 1$. Throughout the paper, we use the notations

$$b^F := F(e_1), \quad \mu_r^F = F(e_1 - b^F e_0)^r, r \in \mathbb{N}.$$

Let $B_2[0, \infty)$ be the set of all functions f defined on the positive real line with some constant $C(f)$ depending only on f, satisfying the condition $|f(x)| \le C(f)(1 + x^2)$. Let $C_2[0, \infty) = C[0, \infty) \cap B_2[0, \infty)$ and by $\widetilde{C}_2[0, \infty)$, we denote the subspace of all continuous functions $f \in B_2[0, \infty)$ for which

$$\lim_{x \to \infty} |f(x)|(1 + x^2)^{-1} < \infty.$$

The weighted modulus of continuity $\Omega(f, \delta)$ (see [104] and [23]), for each $f \in C_2[0, \infty)$ is defined as

$$\Omega(f, \delta) = \sup_{|h|<\delta,\, x\in\mathbb{R}^+} |f(x+h) - f(x)|\left(1 + h^2 + x^2 + h^2 x^2\right)^{-1}.$$

Let $F : \mathbb{D} \to \mathbb{R}$ be a positive linear functional, where $\mathbb{D}$ is a linear subspace of $C[0, \infty)$ which contains $C_2[0, \infty)$ and the polynomials up to degree 6.

3.2 Estimates with Same Basis

Recently Gupta in [70] established the following general result for difference of two operators having the same basis function.

Theorem 3.4 ([70]) *Let* $f^{(s)} \in C_B[0, \infty)$, $s \in \{0, 1, 2\}$ *and* $x \in [0, \infty)$, *then for* $n \in \mathbb{N}$, *we have*

$$|((U_n - V_n)f)(x)| \le \frac{\alpha(x)}{2}\|f''\| + \frac{(1+\alpha(x))}{2}\omega(f'', \delta_1) + 2\omega(f, \delta_2(x)),$$

where $C_B[0, \infty)$ *is the class of bounded continuous functions defined for* $x \ge 0$,
$\|.\| = \sup_{x\in[0,\infty)} |f(x)| < \infty$,

$$\alpha(x) = \sum_{k=0}^{\infty} v_{n,k}(x)(\mu_2^{F_{n,k}} + \mu_2^{G_{n,k}})$$

and

$$\delta_1^2 = \sum_{k=0}^{\infty} v_{n,k}(x)(\mu_4^{F_{n,k}} + \mu_4^{G_{n,k}}),\ \delta_2^2 = \sum_{k=0}^{\infty} v_{n,k}(x)(b^{F_{n,k}} - b^{G_{n,k}})^2.$$

Corollary 3.1 *If the operators* U_n *and* V_n *preserve linear functions, then under the assumptions of Theorem 3.4, we have*

$$|((U_n - V_n)f)(x)| \le \frac{\alpha(x)}{2}\|f''\| + \frac{(1+\alpha(x))}{2}\omega(f'', \delta_1).$$

In [70] and [74] Gupta presented some examples for the difference of Szász–Mirakyan operators and Lupaş operators, respectively, as well as their variants. Additionally, in [71] the following exact estimate for the difference of Baskakov–Szász (case 5 in (2.6.1)) and Baskakov operators is provided:

Proposition 3.1 *Let* $f^{(s)} \in C_B[0, \infty)$, $s \in \{0, 1, 2\}$ *and* $x \in [0, \infty)$, *then for* $n \in \mathbb{N}$, *we have*

$$|((D_{n,n,\infty} - V_n)f)(x)| \leq \frac{\alpha(x)}{2}\|f''\| + \frac{(1+\alpha(x))}{2}\omega(f'', \delta_1) + 2\omega(f, \delta_2(x)),$$

where

$$\alpha(x) = \frac{nx+1}{n^2}, \delta_1^2(x) = \frac{3x^2n(n+1) + 15nx + 9}{n^4}, \delta_2^2(x) = \frac{1}{n^2}.$$

In terms of higher derivatives of functions, Acu and Rasa [10] estimated the following general result.

Theorem 3.5 (see [10]) *Suppose that* $f \in D(I)$ *with* $f^{(i)} \in C_B(I)$ $(i = 2, 3, 4)$. *Then*

$$|((U_n - V_n)f)(x)| \leq \|f^{(2)}\|\gamma(x) + \|f^{(3)}\|\beta(x) + \|f^{(4)}\|\alpha(x) \quad (x \in I),$$

where

$$\gamma(x) := \sum_{k \in K} |\mu_2^{F_{n,k}} - \mu_2^{G_{n,k}}|v_{n,k}(x),$$

$$\beta(x) := \sum_{k \in K} |\mu_3^{F_{n,k}} - \mu_3^{G_{n,k}}|v_{n,k}(x)$$

and

$$\alpha(x) := \sum_{k \in K} (\mu_4^{F_{n,k}} + \mu_4^{G_{n,k}})v_{n,k}(x).$$

For differences of two operators Aral et al. [23] provided a general estimate on polynomials in weighted spaces. While finding the difference between the variants of Baskakov operators Gupta–Agrawal–Rassias [80] considered the following slightly different version of [23]:

Theorem 3.6 *Let* $f \in C_2[0, \infty)$ *with* $f'' \in \widetilde{C}_2[0, \infty)$. *Then*

$$|((U_n - V_n)f)(x)| \leq \frac{1}{2}\|f''\|_2\beta(x) + 8\Omega\left(f'', \delta_1\right)(1 + \beta(x))$$

$$+16\Omega\left(f, \delta_2\right)(\gamma(x) + 1),$$

where

$$\beta(x) = \sum_{k \in \mathbb{K}} v_{n,k}(x) \left\{ \left(1 + \left(b^{F_{n,k}} \right)^2 \right) \mu_2^{F_{n,k}} + \left(1 + \left(b^{G_{n,k}} \right)^2 \right) \mu_2^{G_{n,k}} \right\},$$

$$\gamma(x) = \sum_{k \in \mathbb{K}} v_{n,k}(x) \left(1 + \left(b^{(FG)_{n,k}} \right)^2 \right),$$

$$\delta_1^4(x) = \sum_{k \in \mathbb{K}} v_{n,k}(x) \left\{ \left(1 + \left(b^{F_{n,k}} \right)^2 \right) \mu_6^{F_{n,k}} + \left(1 + \left(b^{G_{n,k}} \right)^2 \right) \mu_6^{G_{n,k}} \right\}$$

and

$$\delta_2^4(x) = \sum_{k \in \mathbb{K}} v_{n,k}(x) \left(1 + \left(b^{(FG)_{n,k}} \right)^2 \right) \left(b^{F_{n,k}} - b^{G_{n,k}} \right)^4,$$

where we suppose that $\delta_1(x) \le 1$, $\delta_2(x) \le 1$ *and* $b^{(FG)_{n,k}} = min\{b^{F_{n,k}}, b^{G_{n,k}}\}$.

Several examples between Baskakov and their variants have been provided in [80] and [150]. We indicate below only two examples of [80].

As an application of Theorem 3.6, the difference between the Baskakov and the Baskakov–Szász operators (case 5 in (2.6.1)) is given below:

Proposition 3.2 *Let* $f \in C_2[0, \infty)$ *with* $f'' \in \tilde{C}_2[0, \infty)$. *Then*

$$\left| ((V_n - D_{n,n,\infty}) f)(x) \right| \le \frac{1}{2} \|f''\|_2 \beta(x) + 8\Omega \left(f'', \delta_1 \right) (1 + \beta(x))$$
$$+ 16\Omega \left(f, \delta_2 \right) (1 + \gamma(x)),$$

where

$$\beta(x) = \frac{1}{n^4} \left[n^3(x^3 + x) + n^2(3x^3 + 6x^2 + 1) + nx(2x^2 + 6x + 7) + 1 \right],$$

$$\delta_1^4(x) = \frac{1}{8n^8} \left[n^5 x^3 (x^2 + 120) + 10n^4 x^2 (x^3 + 165x^2 + 36x + 176) + 5n^3 x (7x^4 \right.$$

$$+ 1980x^3 + 3285x^2 + 352x + 984) + 5n^2 (10x^5 + 3630x^4 + 9711x^3$$

$$+ 8155x^2 + 424) + nx(24x^4 + 9900x^3 + 32370x^2$$

$$\left. + 40775x + 25921) + 2120 \right],$$

$$\gamma(x) = \frac{x^2(n+1) + x + n}{n}, \quad \delta_2^4(x) = \frac{1}{n^5} \left[(n+1)x^2 + x + n \right].$$

The Baskakov–Kantorovich operators for $x \in [0, \infty)$ are defined in (2.2.3) as:

$$(K_n^{\hat{v}} f)(x) = n \sum_{k=0}^{\infty} \binom{n+k-1}{k} \frac{x^k}{(1+x)^{n+k}} \int_{k/n}^{(k+1)/n} f(t)dt.$$

Another application of Theorem 3.6, is the difference between the Baskakov and the Baskakov–Kantorovich operators (2.2.3) provided below:

Proposition 3.3 *Let $f \in C_2[0, \infty)$ with $f'' \in \tilde{C}_2[0, \infty)$. Then*

$$\left|((V_n - K_n^{\hat{v}})f)(x)\right| \leq \frac{1}{2}\|f''\|_2 \beta(x) + 8\Omega\left(f'', \delta_1\right)(1 + \beta(x))$$
$$+ 16\Omega(f, \delta_2)(1 + \gamma(x)),$$

where

$$\beta(x) = \frac{1}{48n^4}\left[4n^2(x^2+1) + 4n(x^2+2x) + 1\right],$$

$$\delta_1^4(x) = \frac{1}{1792n^8}\left[4n^2(x^2+1) + 4n(x^2+2x) + 1\right],$$

$$\delta_2^4(x) = \frac{n(x^2+1) + (x^2+x)}{16n^5},$$

and $\gamma(x)$ is same as in Proposition 3.2.

3.3 Estimates with Different Basis

Gupta and Acu [76] estimated the difference between two operators having different basis as presented in the following theorem:

Theorem 3.7 *If $f \in D(I)$ with $f'' \in C_B(I)$, then*

$$|((U_n - W_n)f)(x)| \leq \alpha(x)\|f''\| + 2\omega_1(f; \delta_1(x)) + 2\omega_1(f; \delta_2(x)), \qquad (3.3.1)$$

where

$$\alpha(x) = \frac{1}{2}\sum_{k \in K}\left(v_{n,k}(x)\mu_2^{F_{n,k}} + u_{n,k}(x)\mu_2^{H_{n,k}}\right),$$

$$\delta_1^2(x) = \sum_{k \in K} v_{n,k}(x)\left(b^{F_{n,k}} - x\right)^2, \quad \delta_2^2(x) = \sum_{k \in K} u_{n,k}(x)\left(b^{H_{n,k}} - x\right)^2.$$

Remark 3.1 If we use a result of Shisha and Mond [149], we can write

$$|((U_n - W_n)f)(x)| \leq |(U_nf)(x) - f(x)| + |(V_nf)(x) - f(x)|$$
$$\leq 2\omega_1(f, v_1(x)) + 2\omega_1(f, v_2(x)),$$

where

$$v_1^2(x) = \left(U_n(e_1 - x)^2\right)(x) = \sum_{k \in K} v_{n,k}(x) F_{n,k}\left((e_1 - x)^2; x\right),$$

$$v_2^2(x) = \left(W_n(e_1 - x)^2\right)(x) = \sum_{k \in K} u_{n,k}(x) H_{n,k}\left((e_1 - x)^2; x\right).$$

Since $F_{n,k}^2(e_1) \leq F_{n,k}(e_1^2)$ and $H_{n,k}^2(e_1) \leq H_{n,k}(e_1^2)$, it follows that $\delta_i(x) \leq v_i(x)$, $i = 1, 2$.

Remark 3.2 If $F_{n,k}^2(e_1) = F_{n,k}(e_1^2)$ and $H_{n,k}^2(e_1) = H_{n,k}(e_1^2)$, then the relation (3.3.1) becomes

$$|((U_n - W_n)f)(x)| \leq 2\omega_1(f, v_1(x))' + 2\omega_1(f, v_2(x)),$$

where v_1, v_2 are defined in Remark 3.1.

The quantitative estimates for the differences of Szász operators and Baskakov operators is provided in the following application of Theorem 3.7:

Proposition 3.4 *Let $f \in C_B[0, \infty)$ and $x \in [0, \infty)$. Then for $n \in \mathbb{N}$, we have*

$$|((V_n - S_n)f)(x)| \leq 2\omega_1\left(f, \sqrt{\frac{x(1+x)}{n}}\right) + 2\omega_1\left(f, \sqrt{\frac{x}{n}}\right).$$

The following proposition provides an estimate for the difference of Baskakov and an integral modification of Szász–Baskakov type operators (see case (6) of (2.6.1)), which constitutes an application of Theorem 3.7:

Proposition 3.5 *Let $f^{(s)} \in C_B[0, \infty), s \in \{0, 1, 2\}$ and $x \in [0, \infty)$, then for $n \in \mathbb{N}, n > 3$, we have*

$$|((D_{n,\infty,n} - V_n)f)(x)| \leq \frac{n^2x^2 + nx + n^2x + n - 1}{2(n-2)^2(n-3)}\|f''\|$$

$$+2\omega_1\left(f; \sqrt{\frac{x(1+x)}{n}}\right)$$

$$+2\omega_1\left(f; \frac{\sqrt{4x^2 + (4+n)x + 1}}{n-2}\right).$$

Theorem 3.8 *Let* $I = [0, 1]$, $f \in C[0, 1]$, $0 < h \le \dfrac{1}{2}$, $x \in [0, 1]$. *Then*

$$|((U_n - W_n)f)(x)| \le \frac{3}{2}\left(1 + \frac{\alpha(x)}{h^2}\right)\omega_2(f, h) + (\delta_1(x) + \delta_2(x))\frac{10}{h}\omega_1(f, h).$$

Păltănea [139] as well as Păltănea and Gonska [57] proposed a new class of Bernstein–Durrmeyer type operators $U_n^\rho : C[0, 1] \to \prod_n$, defined as

$$(U_n^\rho f)(x) = \sum_{k=1}^{n-1}\left(\int_0^1 \frac{t^{k\rho-1}(1-t)^{(n-k)\rho-1}}{B(k\rho, (n-k)\rho)} f(t)dt\right) p_{n,k}(x)$$
$$+ f(0)(1-x)^n + f(1)x^n,$$

where $p_{n,k}(x) = \binom{n}{k}x^k(1-x)^{n-k}$. These operators constitute a link between the genuine Bernstein–Durrmeyer operators U_n ($\rho = 1$) and the classical Bernstein operators.

Stancu [152] introduced a sequence of positive linear operators $P_n^{<\alpha>} : C[0, 1] \to C[0, 1]$, depending on a parameter $\alpha \ge 0$ as follows

$$(P_n^{<\alpha>}f)(x) = \sum_{k=0}^{n} f\left(\frac{k}{n}\right) p_{n,k}^{<\alpha>}(x), \quad x \in [0, 1],$$

where

$$p_{n,k}^{<\alpha>}(x) = \binom{n}{k}\frac{x^{[k,-\alpha]}(1-x)^{[n-k,-\alpha]}}{1^{[n,-\alpha]}}$$

and

$$x^{[n,\alpha]} := x(x - \alpha) \cdots (x - \overline{n-1}\alpha).$$

As a special case when $\alpha = 0$, we obtain the Bernstein operators. The case $\alpha = \dfrac{1}{n}$ of these operators was considered by L. Lupaş and A. Lupaş [116] as follows

$$(P_n^{<\frac{1}{n}>}f)(x) = \frac{2n!}{(2n)!}\sum_{k=0}^{n}\binom{n}{k}f\left(\frac{k}{n}\right)(nx)_k(n-nx)_{n-k}.$$

Neer and Agrawal [134] introduced a genuine-Durrmeyer type modification of these operators as presented below:

$$(\tilde{U}_n^\rho f)(x) = \sum_{k=0}^{n} p_{n,k}^{<\frac{1}{n}>}(x)\int_0^1 \frac{t^{k\rho-1}(1-t)^{(n-k)\rho-1}}{B(k\rho, (n-k)\rho)} f(t)dt, \quad \rho > 0, \ f \in C[0, 1].$$

An application of Theorem 3.8 is the following:

Proposition 3.6 *The following inequalities hold*

$$\left|((U_n^\rho - \tilde{U}_n^\rho)f)(x)\right| \le \frac{x(1-x)}{n\rho + 1}\|f''\| + 2\omega_1\left(f, \sqrt{\frac{x(1-x)}{n}}\right)$$

$$+ 2\omega_1\left(f, \sqrt{\frac{2x(1-x)}{n+1}}\right), \quad f'' \in C[0,1]$$

$$\left|((U_n^\rho - \tilde{U}_n^\rho)f)(x)\right| \le 3\omega_2\left(f, \sqrt{\frac{x(1-x)}{n\rho + 1}}\right)$$

$$+ 10(1+\sqrt{2})\sqrt{\frac{n\rho + 1}{n}}\,\omega_1\left(f, \sqrt{\frac{x(1-x)}{n\rho + 1}}\right), \quad f \in C[0,1].$$

The Durrmeyer type modification introduced by Gupta et al. [92] is the following

$$(D_n^{<\frac{1}{n}>}f)(x) = (n+1)\sum_{k=0}^{n} p_{n,k}^{<\frac{1}{n}>}\int_0^1 p_{n,k}(t)f(t)dt, \quad f \in C[0,1].$$

As an application of Theorem 3.8, the difference between $D_n^{<\frac{1}{n}>}$ and the usual Durrmeyer operators $D_{n,-n,-n}$ (see case (4) of (2.6.1)) is presented below:

Proposition 3.7 *The following inequalities hold:*

(i) For $f'' \in C[0,1]$, we have

$$\left|((D_{n,-n,-n} - D_n^{<\frac{1}{n}>})f)(x)\right| \le \frac{n(2n+1)(n-1)x(1-x) + 2(n+1)^2}{2(n+1)(n+3)(n+2)^2}\|f''\|$$

$$+ 2\omega_1\left(f, \sqrt{\frac{nx(1-x) + (2x-1)^2}{(n+2)^2}}\right)$$

$$+ 2\omega_1\left(f, \sqrt{\frac{2n^2x(1-x) + (2x-1)^2(n+1)}{(n+2)^2(n+1)}}\right).$$

(ii) For $f \in C[0, 1]$, we have

$$\left|((D_{n,-n,-n} - D_n^{<\frac{1}{n}>})f)(x)\right|$$

$$\leq 3\omega_2\left(f, \sqrt{\frac{n(2n+1)(n-1)x(1-x)+2(n+1)^2}{2(n+1)(n+3)(n+2)^2}}\right)$$

$$+20\sqrt{3}\omega_1\left(f, \sqrt{\frac{n(2n+1)(n-1)x(1-x)+2(n+1)^2}{2(n+1)(n+3)(n+2)^2}}\right).$$

Very recently, Gupta et al. [77] provided estimates for the differences between some of the most representative operators. They proved the following general result:

Theorem 3.9 ([77]) *Let $f \in D(I)$. If $f^{(i)} \in C_B(I)$ $(i = 2, 3, 4)$, then*

$$|((U_n - W_n)f)(x)| \leq A(x)\|f^{(4)}\| + B(x)\|f^{(3)}\| + C(x)\|f^{(2)}\|$$
$$+ 2\omega_1(f, \delta_1(x)) + 2\omega_1(f, \delta_2(x)) \quad (x \in I),$$

where $\omega_1(f, \cdot)$ is the usual modulus of continuity,

$$A(x) = \frac{1}{4!}\sum_{k\in K}(v_{n,k}(x)\mu_4^{F_{n,k}} + u_{n,k}(x)\mu_4^{H_{n,k}}),$$

$$B(x) = \frac{1}{3!}\left|\sum_{k\in K}v_{n,k}(x)\mu_3^{F_{n,k}} - \sum_{k\in K}u_{n,k}(x)\mu_3^{H_{n,k}}\right|,$$

$$C(x) = \frac{1}{2!}\left|\sum_{k\in K}v_{n,k}(x)\mu_2^{F_{n,k}} - \sum_{k\in K}u_{n,k}(x)\mu_2^{H_{n,k}}\right|,$$

$$\delta_1(x) = \left(\sum_{k\in K}v_{n,k}(x)\left(b^{F_{n,k}} - x\right)^2\right)^{1/2}$$

and

$$\delta_2(x) = \left(\sum_{k\in K}u_{n,k}(x)\left(b^{H_{n,k}} - x\right)^2\right)^{1/2}.$$

They presented several applications of this theorem and we mention below some of them. The difference between the Baskakov and the Szász–Baskakov operators (see case (6) of (2.6.1)) is given by:

Proposition 3.8 *If $f \in D([0, \infty))$ with $f^{(i)} \in C_B[0, \infty)$ $(i = 2, 3, 4)$, then for each $x \in [0, \infty)$, it is asserted that*

$$|((D_{n,\infty,n} - V_n)f)(x)| \le A(x)\|f^{(4)}\| + B(x)\|f^{(3)}\| + C(x)\|f^{(2)}\|$$
$$+ 2\omega_1(f, \delta_1(x)) + 2\omega_1(f, \delta_2(x)),$$

where

$$A(x) = \frac{1}{8(n-5)(n-4)(n-3)(n-2)^4}\left\{x^2(x+1)^2 n^5\right.$$
$$+ x(4x^3 + 14x^2 + 14x + 5)n^4$$
$$\left. + (x+1)(24x^2 + 5x + 3)n^3 + 28x^2 + 7x - 8)n^2\right\},$$
$$B(x) = \frac{x(x+1)(2x+1)n^3 + (2x+1)(3x+1)n^2 - n}{3(n-2)^2(n-3)(n-4)},$$
$$C(x) = \frac{x(1+x)n^2 + (x+1)n - 1}{2(n-2)^2(n-3)},$$
$$\delta_1(x) = \sqrt{\frac{x(1+x)}{n}}, \quad \delta_2(x) = \frac{\sqrt{4x^2 + (4+n)x + 1}}{(n-2)}.$$

The Szász–Mirakyan–Kantorovich operators are defined in (2.2.4) by

$$(K_n^S f)(x) = n \sum_{k=0}^{\infty} s_k(nx) \int_{k/n}^{(k+1)/n} f(t)dt.$$

The difference between the Baskakov and the Szász–Mirakyan–Kantorovich operators is provided below:

Proposition 3.9 *Let $I = [0, \infty)$. If $f \in D(I)$ with $f^{(i)} \in E_B(I)$ $(i = 2, 3, 4)$, then for each $x \in [0, \infty)$, it is asserted that*

$$|((K_n^S - V_n)f)(x)| \le A(x)\|f^{(4)}\| + C(x)\|f^{(2)}\| + 2\omega_1(f, \delta_1) + 2\omega_1(f, \delta_2),$$

where

$$A(x) = \frac{1}{1920n^4} \quad and \quad C(x) = \frac{1}{24n^2}$$

and

$$\delta_1(x) = \sqrt{\frac{x(1+x)}{n}} \quad and \quad \delta_2(x) = \frac{\sqrt{4nx + 1}}{2n}.$$

The difference between two genuine-Durrmeyer type operators (defined in Proposition 3.6) is provided below:

Proposition 3.10 Let $f \in C^4[0, 1]$. Then the following inequality holds true:

$$\left|\left(U_m^\rho - \tilde{U}_m^\rho f\right)(x)\right| \leq A(x)\|f^{(4)}\| + B(x)\|f^{(3)}\| + C(x)\|f^{(2)}\|$$

$$+ 2\omega_1\left(f, \delta_1(x)\right) + 2\omega_1\left(f, \delta_2(x)\right),$$

where

$$A(x) := \frac{x(1-x)(n-1)}{8m^3(m\rho+1)(m\rho+2)(m\rho+3)(m+1)(m+2)(m+3)}$$

$$\cdot \left\{m\rho(3m^4 + 5m^3 + 7m^2 - 5m - 6) + 4m^5 + 4m^4 + 4m^3 - 30m^2\right.$$

$$+ 30m+36+x(1-x)(m-2)(m-3)(m\rho-6)(2m^3+6m^2+11m+6),$$

$$B(x) := \frac{x(1-x)|1-2x|(m-2)(m-1)(3m+2)}{3(m\rho+1)(m\rho+2)m^2(m+1)(m+2)},$$

$$C(x) := \frac{x(1-x)(m-1)}{2(m\rho+1)m(m+1)},$$

$$\delta_1(x) := \sqrt{\frac{x(1-x)}{m}}, \quad \delta_2(x) := \sqrt{\frac{2x(1-x)}{m+1}}.$$

The main result for the difference of operators having different basis in the weighted space was established by Gupta [69] as the following theorem:

Theorem 3.10 ([69]) Let $f \in C_2[0, \infty)$ with $f'' \in \tilde{C}_2[0, \infty)$. Then

$$|((U_n - W_n)f)(x)| \leq \frac{1}{2}\|f''\|_2(\beta_1(x) + \beta_2(x))$$

$$+ 8\Omega\left(f'', \delta_1\right)(1 + \beta_1(x)) + 8\Omega\left(f'', \delta_2\right)(1 + \beta_2(x))$$

$$+ 16\Omega(f, \delta_3)(\gamma_1(x)+1) + 16\Omega(f, \delta_4)(\gamma_2(x)+1),$$

where

$$\beta_1(x) = \sum_{k\in\mathbb{K}} v_{n,k}(x)\left(1 + \left(b^{F_{n,k}}\right)^2\right)\mu_2^{F_{n,k}},$$

$$\beta_2(x) = \sum_{k\in\mathbb{K}} u_{n,k}(x)\left(1 + \left(b^{H_{n,k}}\right)^2\right)\mu_2^{H_{n,k}},$$

$$\delta_1^4(x) = \sum_{k\in\mathbb{K}} v_{n,k}(x)\left(1 + \left(b^{F_{n,k}}\right)^2\right)\mu_6^{F_{n,k}},$$

$$\delta_2^4(x) = \sum_{k\in\mathbb{K}} u_{n,k}(x) \left(1 + \left(b^{H_{n,k}}\right)^2\right) \mu_6^{H_{n,k}},$$

$$\delta_3^4(x) = \sum_{k\in\mathbb{K}} v_{n,k}(x) \left(1 + \left(b^{F_{n,k}}\right)^2\right) (b^{F_{n,k}} - x)^4,$$

$$\delta_4^4(x) = \sum_{k\in\mathbb{K}} u_{n,k}(x) \left(1 + \left(b^{H_{n,k}}\right)^2\right) (b^{H_{n,k}} - x)^4.$$

$$\gamma_1(x) = \sum_{k\in\mathbb{K}} v_{n,k}(x)(1 + (b^{F_{n,k}})^2),$$

$$\gamma_2(x) = \sum_{k\in\mathbb{K}} u_{n,k}(x)(1 + (b^{H_{n,k}})^2).$$

We suppose that $\delta_1(x) \le 1, \delta_2(x) \le 1, \delta_3(x) \le 1, \delta_4(x) \le 1$.

In [69] the authors considered several examples of Theorem 3.10.
The difference estimate between the Baskakov and the Szász–Mirakyan operators is provided below:

Proposition 3.11 *Let* $f \in C_2[0, \infty)$ *with* $f'' \in \tilde{C}_2[0, \infty)$. *Then we have*

$$|((V_n - S_n)f)(x)| \le 16\Omega(f, \delta_3)(\gamma_1(x) + 1) + 16\Omega(f, \delta_4)(\gamma_2(x) + 1),$$

where

$$\delta_3^4(x) = \frac{3x^4}{n^2} + \frac{36x^3}{n^3} + \frac{30x^2}{n^4} + \frac{x}{n^5} + \frac{x}{n^3},$$

$$\delta_4^4(x) = \frac{x(x+1)}{n^5}\Big[3n^3x(x+1)(x^2+1) + n^2(61x^4 + 96x^3 + 42x^2 + 6x + 1)$$

$$+ nx(178x^3 + 332x^2 + 183x + 27)$$

$$+ (120x^4 + 240x^3 + 150x^2 + 30x + 1)\Big],$$

$$\gamma_1(x) = 1 + x^2 + \frac{x}{n}, \quad \gamma_2(x) = 1 + x^2 + \frac{x(1+x)}{n}$$

and where we suppose that $\delta_3(x) \le 1, \delta_4(x) \le 1$.

3.4 Differences in Terms of Weighted Modulus $\omega_\varphi(f, h)$

The weighted modulus $\omega_\varphi(f, h)$ introduced by Păltănea in [137] is defined as

$$\omega_\varphi(f, h) = \sup\left\{|f(x) - f(y)| : x \ge 0, y \ge 0, |x - y| \le h\varphi\left(\frac{x+y}{2}\right)\right\}, h \ge 0,$$

where $\varphi(x) = \frac{\sqrt{x}}{1+x^m}$, $x \in [0, \infty)$, $m \in \mathbb{N}$, $m \geq 2$. We consider here those functions, for which we have the property

$$\lim_{h \to 0} \omega_\varphi(f, h) = 0.$$

It is easy to verify that this property is satisfied for f being an algebraic polynomial of degree $\leq m$. Following Theorem 2 of [137], the function f satisfies the following two conditions:

- The function $f \circ e_2$ is uniformly continuous on $[0, 1]$
- The function $f \circ e_v$, $v = \frac{2}{2m+1}$ is uniformly continuous on $[1, \infty)$, where $e_v(x) = x^v$, $x \geq 0$

We denote by $W_\varphi[0, \infty)$ the subspace of all real functions defined on $[0, \infty)$, satisfying the above conditions.

Instead of this, Gupta and Tachev in [98] studied the difference of two operators with different basis functions of discrete operators and furthermore the arbitrary positive linear operators, including the integral representation. The only information required is a good (exact if possible) representation of moments of the two operators M_n and L_n of order 6.

In our note we consider l.p.o. $L_n : E \to C[0, \infty)$, where E is a subspace of $C[0, \infty)$, such that $C_k[0, \infty) \subset E$, with $k = \max\{m + r + 1, 2r + 2, 2m\}$, $r \in \mathbb{N}$ and

$$C_k[0, \infty) := \{f \in C[0, \infty), \exists M > 0 : |f(x)| \leq M(1 + x^k), \forall x \geq 0, k \in \mathbb{N}\}.$$

Let $\mu_{n,m}^L(x)$, $m \in \mathbb{N}$ be the moment of order m of L_n, i.e.

$$\mu_{n,m}^L(x) = L_n((t - x)^m, x).$$

Theorem 3.11 ([98]) *Let* $L_n, M_n : E \to C[0, \infty)$, $C_k[0, \infty) \subset E$, $k = \max\{m + 3, 6, 2m\}$ *be two sequences of linear positive operators. If* $f \in C^2[0, \infty) \cap E$ *and* $f'' \in W_\varphi[0, \infty)$, *then we have for* $x \in (0, \infty)$ *that*

$$|(L_n f)(x) - (M_n f)(x)|$$

$$\leq |f'(x)| \cdot \left| \mu_{n,1}^{L_n}(x) - \mu_{n,1}^{M_n}(x) \right|$$

$$+ \frac{1}{2}|f''(x)| \cdot \left| \mu_{n,2}^{L_n}(x) - \mu_{n,2}^{M_n}(x) \right|$$

$$+ \frac{1}{2} \left[\mu_{n,2}^{L_n}(x) + \sqrt{2} \cdot \sqrt{L_n \left(\left[1 + \left(x + \frac{|t - x|}{2} \right)^m \right] \right)} \right] \omega_\varphi \left(f'', \left(\frac{\mu_{n,6}^{L_n}}{x} \right)^{1/2} \right)$$

$$+ \frac{1}{2} \left[\mu_{n,2}^{M_n}(x) + \sqrt{2} \cdot \sqrt{M_n \left(\left[1 + \left(x + \frac{|t - x|}{2} \right)^m \right] \right)} \right] \omega_\varphi \left(f'', \left(\frac{\mu_{n,6}^{M_n}}{x} \right)^{1/2} \right).$$

Remark 3.3 If both operators L_n and M_n reproduce linear functions, we have $\mu_{n,1}^{L_n}x = \mu_{n,1}^{M_n}x = 0$. Therefore we can omit the summand containing $f'(x)$. Thus in all exponential operators as indicated in Table 1.1.14, the term containing $f'(x)$ is absent.

The following example provides the difference between Baskakov operators V_n and Szász–Mirakyan operators S_n

Example 3.1 Let $S_n, V_n : E \to C[0, \infty)$, $C_k[0, \infty) \subset E$, $k = \max\{m + 3, 6, 2m\}$ be two sequences of linear positive operators. If $f \in C^2[0, \infty) \cap E$ and $f'' \in W_\varphi[0, \infty)$, then we have for $x \in (0, \infty)$ that

$$|(S_n f)(x) - (V_n f)(x)|$$

$$\leq \frac{x^2}{2n}|f''(x)| + \frac{1}{2}\left[\frac{x}{n} + \sqrt{2A_{n,m,x}}\right]\omega_\varphi\left(f'', \sqrt{\frac{1}{n^5} + \frac{25x}{n^4} + \frac{15x^2}{n^3}}\right)$$

$$+\frac{1}{2}\left[\frac{x(1+x)}{n} + \sqrt{2V_{n,m,x}}\right]$$

$$\times \omega_\varphi\left(f'', \left(\frac{1 + 31x + 180x^2 + 390x^3 + 360x^4 + 120x^5}{n^5}\right.\right.$$

$$+\frac{25x + 288x^2 + 667x^3 + 534x^4 + 130x^5}{n^4}$$

$$\left.\left.+\frac{15x^2 + 105x^3 + 105x^4 + 15x^5}{n^3}\right)^{1/2}\right),$$

where

$$A_{n,m,x} = S_n\left(\left[1 + \left(x + \frac{|t - x|}{2}\right)^m\right]^2, x\right),$$

$$V_{n,m,x} = V_n\left(\left[1 + \left(x + \frac{|t - x|}{2}\right)^m\right]^2, x\right).$$

The well-known Phillips operators [141] are defined as

$$(\widetilde{P}_n f)(x) = n\sum_{k=1}^{\infty} e^{-nx}\frac{(nx)^k}{k!}\int_0^\infty e^{-nt}\frac{(nt)^{k-1}}{(k-1)!}f(t)dt + e^{-nx}f(0).$$

The Lupaş operators are defined as

$$(\widetilde{U}_n f)(x) := \sum_{k=0}^{\infty} 2^{-nx} \frac{(nx)_k}{k!\, 2^k}\, f\left(\frac{k}{n}\right).$$

We present now the following quantitative estimate in the form of an application of Theorem 3.11 for the difference between Szász–Mirakyan and Phillips operators.

Example 3.2 Let $S_n, \widetilde{P}_n : E \to C[0, \infty)$, $C_k[0, \infty) \subset E$, $k = \max\{m+3, 6, 2m\}$ be two sequences of linear positive operators. If $f \in C^2[0, \infty) \cap E$ and $f'' \in W_\varphi[0, \infty)$, then we have for $x \in (0, \infty)$ that

$$\left|(S_n f)(x) - (\widetilde{P}_n f)(x)\right|$$

$$\leq \frac{x}{2n}|f''(x)| + \frac{1}{2}\left[\frac{x}{n} + \sqrt{2A_{n,m,x}}\right]\omega_\varphi\left(f'', \sqrt{\frac{1}{n^5} + \frac{25x}{n^4} + \frac{15x^2}{n^3}}\right)$$

$$+\frac{1}{2}\left[\frac{2x}{n} + \sqrt{2C_{n,m,x}}\right]\omega_\varphi\left(f'', \sqrt{\frac{720}{n^5} + \frac{576x}{n^4} + \frac{432}{n^4} + \frac{120x^2}{n^3}}\right),$$

where

$$A_{n,m,x} = S_n\left(\left[1 + \left(x + \frac{|t-x|}{2}\right)^m\right]^2, x\right),$$

$$C_{n,m,x} = \widetilde{P}_n\left(\left[1 + \left(x + \frac{|t-x|}{2}\right)^m\right]^2, x\right).$$

The following quantitative estimate is application of Theorem 3.11 for the difference of Phillips and Lupaş operators.

Example 3.3 Let $\widetilde{P}_n, \widetilde{U}_n : E \to C[0, \infty)$, $C_k[0, \infty) \subset E$, $k = \max\{m+3, 6, 2m\}$ be two sequences of linear positive operators. If $f \in C^2[0, \infty) \cap E$ and $f'' \in W_\varphi[0, \infty)$, then we have for $x \in (0, \infty)$ that

$$\left|(\widetilde{P}_n f)(x) - (\widetilde{U}_n f)(x)\right|$$

$$\leq \frac{1}{2}\left[\frac{2x}{n} + \sqrt{2C_{n,m,x}}\right]\omega_\varphi\left(f'', \sqrt{\frac{720}{n^5} + \frac{576x}{n^4} + \frac{432}{n^4} + \frac{120x^2}{n^3}}\right)$$

$$+\frac{1}{2}\left[\frac{2x}{n} + \sqrt{2D_{n,m,x}}\right]\omega_\varphi\left(f'', \sqrt{\frac{1082 + 1140nx + 120n^2x^2}{n^5}}\right),$$

where

$$D_{n,m,x} = \tilde{U}_n\left(\left[1 + \left(x + \frac{|t-x|}{2}\right)^m\right]^2, x\right),$$

$$C_{n,m,x} = \tilde{P}_n\left(\left[1 + \left(x + \frac{|t-x|}{2}\right)^m\right]^2, x\right).$$

The following quantitative estimate constitutes an application of Theorem 3.11 for the difference of Phillips and Cismaşiu operators.

Example 3.4 Let $\tilde{P}_n, C_n : E \to C[0, \infty)$, $C_k[0, \infty) \subset E$, $k = \max\{m+3, 6, 2m\}$ be two sequences of linear positive operators. If $f \in C^2[0, \infty) \cap E$ and $f'' \in W_\varphi[0, \infty)$, then we have for $x \in (0, \infty)$ that

$$\left|(\tilde{P}_n f)(x) - (C_n f)(x)\right|$$

$$\leq \frac{|2x(x-1)|}{n}|f''(x)| + \frac{1}{2}\left[\frac{2x}{n} + \sqrt{2C_{n,m,x}}\right]$$

$$\times \omega_\varphi\left(f'', \sqrt{\frac{720}{n^5} + \frac{576x}{n^4} + \frac{432}{n^4} + \frac{120x^2}{n^3}}\right)$$

$$+ \frac{1}{2}\left[\frac{2x^2}{n} + \sqrt{2E_{n,m,x}}\right]\omega_\varphi\left(f'', \sqrt{\frac{40(96 + 52n + 3n^2)x^5}{n^5}}\right),$$

where

$$E_{n,m,x} = C_n\left(\left[1 + \left(x + \frac{|t-x|}{2}\right)^m\right]^2, x\right),$$

$$C_{n,m,x} = \tilde{P}_n\left(\left[1 + \left(x + \frac{|t-x|}{2}\right)^m\right]^2, x\right).$$

3.5 Difference and Derivatives

Recently Acu and Rasa [9] as well as Acu et al. [8] estimated the differences of certain positive linear operators (defined on bounded or unbounded intervals) and their derivatives, in terms of the modulus of continuity. In this section we present some of the most important results of recent studies.

For the well-known Bernstein polynomials, using the representation

$$(B_n f)^{(r)} = (n)_r \sum_{i=0}^{n-r} p_{n-r,i}(x) \Delta_{1/n}^r f\left(\frac{i}{n}\right),$$

the following estimate was established by Acu-Rasa:

Theorem 3.12 ([9]) *For Bernstein operators, the following property holds:*

$$\left\| (B_n f)^{(r)} - B_{n-r}\left(f^{(r)}\right) \right\| \leq \frac{r(r-1)}{2n} \|f^{(r)}\| + \omega\left(f^{(r)}, \frac{r}{n}\right),$$

where $\|.\|$ denotes the supremum-norm.

For $f \in L_1[0, 1]$, the Bernstein–Kantorovich operators (2.2.1) are defined as

$$(K_n^B f)(x) = (n+1) \sum_{k=0}^{n} p_{n,k}(x) \int_{k/(n+1)}^{(k+1)/(n+1)} f(t)dt,$$

where the Bernstein basis is $p_{n,k}(x) = \binom{n}{k} x^k (1-x)^{n-k}$. Clearly

$$(\widehat{K}_n f)(x) = [(B_{n+1}F)(x)]' \text{ where } F(x) = \int_0^x f(t)dt.$$

Theorem 3.13 ([9]) *For Bernstein–Kantorovich operators, the following property holds:*

$$\left\| \left(K_n^B f\right)^{(r)} - K_{n-r}^B\left(f^{(r)}\right) \right\| \leq \frac{r(r+1)}{2(n+1)} \|f^{(r)}\| + \omega\left(f^{(r)}, \frac{r+1}{n+1}\right),$$

where $f \in C^r[0, 1], r = 0, 1, 2, \ldots n$ and $\|.\|$ denotes the supremum-norm.

Let $w^{(\alpha,\beta)}(x) = x^\alpha (1-x)^\beta, \alpha, \beta > -1$ be a Jacobi weight function on the interval $(0, 1)$ and $L_p^{w^{(\alpha,\beta)}}[0, 1]$ denote the space of Lebesgue-measurable functions f on $[0, 1]$ for which the weighted L_p-norm is finite. The Durrmeyer operators can be generalized as follows:

$$(M_n^{(\alpha,\beta)} f)(x) = \sum_{k=0}^{n} p_{n,k}(x) \frac{1}{c_{n,k}^{(\alpha,\beta)}} \int_0^1 p_{n,k}(t) w^{(\alpha,\beta)}(t) f(t) dt$$

As a special case for $\alpha = \beta = 0$, we obtain the Bernstein–Durrmeyer operators.

Theorem 3.14 ([9]) *For Bernstein–Durrmeyer operators with Jacobi weights, the following property holds:*

$$\left\| \frac{\Gamma(n+\alpha+\beta+r+2)(n-r+1)}{\Gamma(n+\alpha+\beta+2)(n+1)} \left(M_n^{(\alpha,\beta)} f \right)^{(r)} - M_{n-r}^{(\alpha,\beta)} \left(f^{(r)} \right) \right\|$$

$$\leq \frac{1}{4} \| f^{(r+2)} \| \frac{(n+\alpha+\beta+3)}{(n+\alpha+\beta+3)^2 - r^2} + \omega \left(f^{(r)}, \frac{r(n-r+|\beta-\alpha|)}{(n+\alpha+\beta+2)^2 - r^2} \right),$$

where $f \in C^{r+2}[0,1], r = 0, 1, 2, \ldots n$ and $\|.\|$ denotes the supremum-norm.

Additionally, equivalent estimates for the Bernstein–Durrmeyer operators with Jacobi weights along with the quantitative estimate for the genuine Bernstein–Durrmeyer operators have been discussed in [9].

Let $c \in \mathbb{R}, n \in \mathbb{R}, n > c$ for $c \geq 0$ and $-n/c \in \mathbb{N}$ for $c < 0$. Furthermore let $I_c = [0, \infty)$ for $c \geq 0$ and $I_c = [0, -1/c]$ for $c < 0$. Consider $f : I_c \longrightarrow \mathbb{R}$ given in such a way that the corresponding integrals and series are convergent. The Baskakov type operators are defined as follows (see [120])

$$(\mathcal{B}_{n,c} f)(x) = \sum_{j=0}^{\infty} p_{n,j}^{[c]}(x) f \left(\frac{j}{n} \right),$$

where

$$p_{n,j}^{[c]}(x) = \begin{cases} \dfrac{n^j}{j!} x^j e^{-nx} & , c = 0, \\[3mm] \dfrac{n^{c,\bar{j}}}{j!} x^j (1+cx)^{-\left(\frac{n}{c}+j\right)} & , c \neq 0, \end{cases}$$

and $a^{c,\bar{j}} := \prod_{l=0}^{j-1}(a+cl), \quad a^{c,\bar{0}} := 1$.

Denote by $V_n := \mathcal{B}_{n,1}$ the classical Baskakov operators, which for $x \in [0, \infty)$ are defined as follows:

$$V_n(f,x) := \sum_{k=0}^{\infty} f \left(\frac{k}{n} \right) v_{n,k}(x), \text{ where } v_{n,k}(x) := \binom{n+k-1}{k} \frac{x^k}{(1+x)^{n+k}}.$$

Denote

$$V_n^{[r]}(f,x) := \sum_{k=0}^{\infty} f \left(\frac{k}{n} \right) v_{n+r,k}(x).$$

The rth derivative of the Baskakov operators can be written as follows:

$$(V_n^{(r)} f)(x) = \sum_{k=0}^{\infty} \Delta_{\frac{1}{n}}^r f\left(\frac{k}{n}\right) \frac{(n+k+r-1)!}{(n-1)!} (1+x)^{-n-k-r} \frac{x^k}{k!}.$$

Using this representation Acu et al. [8] obtained the following quantitative estimate:

Theorem 3.15 ([8]) *For* $r \geq 0$ *and* $f^{(r)} \in C_B[0, \infty)$, *the Baskakov operators satisfy*

$$\left\| (V_n f)^{(r)} - V_n^{[r]}\left(f^{(r)}\right) \right\| \leq \frac{r! - 1}{n} \| f^{(r)} \| + \omega\left(f^{(r)}, \frac{r}{n}\right).$$

Denote by $S_n := \mathcal{B}_{n,0}$ the classical Szász-Mirakjan operators defined as follows:

$$(S_n f)(x) := e^{-nx} \sum_{j=0}^{\infty} \frac{(nx)^j}{j!} f\left(\frac{j}{n}\right), \quad x \in [0, \infty).$$

The derivative of the Szász-Mirakjan operators can be written as

$$(S_n^{(r)} f)(x) = \sum_{k=0}^{\infty} n^r \Delta_{\frac{1}{n}}^r f\left(\frac{k}{n}\right) \frac{(xn)^k}{k!} e^{-nx}.$$

Using this representation, Acu et al. [8] obtained for the derivatives of the Szász-Mirakjan operators the following quantitative estimate:

Theorem 3.16 ([8]) *For* $r \geq 0$ *and for* $f^{(r)} \in C_b[0, \infty)$, *the Szász-Mirakjan operators satisfy*

$$\left\| (S_n f)^{(r)} - \left(S_n f^{(r)}\right) \right\| \leq \omega\left(f^{(r)}, \frac{r}{n}\right).$$

A generalization of these operators has been studied by López-Moreno in [113] as follows

$$(L_{n,s} f)(x) = \sum_{k=0}^{\infty} (-1)^s f\left(\frac{k}{n}\right) \frac{\phi_n^{(k+s)}(x)}{n^s} \frac{(-x)^k}{k!}, \quad x \in [0, \infty),$$

where $f : [0, \infty) \to \mathbb{R}$, $n \in \mathbb{N}$, and the sequence (ϕ_n) of analytic functions $\phi_n : [0, \infty) \to \mathbb{R}$ verifies the conditions:

(i) $\phi_n(0) = 1$, for every $n \in \mathbb{N}$,
(ii) $(-1)^k \phi_n^{(k)}(x) \geq 0$, for every $n \in \mathbb{N}$, $x \in [0, \infty)$, $k \in \mathbb{N}_0$.

The derivative of the operator $L_{n,s}$ satisfies the following representation (see [113, p.147]):

$$(L_{n,s}^{(r)} f)(x) = (-1)^r \sum_{k=0}^{\infty} (-1)^s \Delta_{\frac{1}{n}}^r f\left(\frac{k}{n}\right) \frac{\phi_n^{(k+s+r)}(x)}{n^s} \frac{(-x)^k}{k!}.$$

Applying such a representation, Acu et al. [8] established the Theorem below:

Theorem 3.17 ([8]) *For* $r \geq 0$ *and* $f^{(r)} \in C_B[0, \infty)$, *the positive linear operators* $L_{n,s}$ *satisfy*

$$\left\| (L_{n,s} f)^{(r)} - \left(L_{n,s+r} f^{(r)} \right) \right\| \leq \left(1 + \mathcal{O}(n^{-1}) \right) \omega\left(f^{(r)}, \frac{r}{n} \right).$$

The Kantorovich modifications of the operators $\mathcal{B}_{n,c}$ are defined for $n > (k+1)c$ by

$$\mathcal{K}_{n,c}^{(k)} := \frac{(n-ck)^k}{(n-ck)^{c,\overline{k}}} D^k \mathcal{B}_{n-ck,c} \mathcal{I}_k,$$

where $a^{c,\overline{j}} := \prod_{l=0}^{j-1} (a + cl), a^{c,\overline{0}} := 1$, D^k denotes the k-th order ordinary differential operator and

$$(\mathcal{I}_k f)(x) = \begin{cases} f, & k = 0 \\ \int_0^x \frac{(x-t)^{k-1}}{(k-1)!} f(t) dt & k \in N. \end{cases}$$

Let $n, c, k \geq 0, n > (k+1)c$, be fixed. Using the well-known representation of $B_{n,c}^{(k)}$ (see [101]), one may write

$$(\mathcal{K}_{n,c}^{(k)} f)(x) = \sum_{j=0}^{\infty} k! p_{n,j}^{[c]}(x) \left[\frac{j}{n-ck}, \frac{j+1}{n-ck}, \dots, \frac{j+k}{n-ck}, \mathcal{I}_k f \right]$$

$$= \sum_{j=0}^{\infty} p_{n,j}^{[c]}(x) f(\xi_j), \quad \frac{j}{n-ck} < \xi_j < \frac{j+k}{n-ck}.$$

The domain of $\mathcal{K}_{n,c}^{(k)}$ is a linear subspace $H_{n,c}^{(k)}$ of $C[0, \infty)$ if $c \geq 0$, or $C[0, -1/c]$ if $c < 0$, containing the polynomial functions. For $j \geq 0$ and $f \in H_{n,c}^{(k)}$ let

$$F_j(f) = k! \left[\frac{j}{n-ck}, \frac{j+1}{n-ck}, \dots, \frac{j+k}{n-ck}, \mathcal{I}_k f \right],$$

$$G_j(f) = f\left(\frac{2j+k}{2(n-ck)} \right).$$

The discrete operators associated with $\mathcal{K}_{n,c}^{(k)}$ are given by

$$(\mathcal{D}_{n,c}^{(k)} f)(x) = \sum_{j=0}^{\infty} p_{n,j}^{[c]}(x) G_j(f).$$

The Peetre's K-functional (see [43, 46]) for $f \in C_B[0, \infty)$ are defined as

$$K_2(f, \lambda) = \inf \left\{ \|f - g\| + \lambda \|g''\| : g, g'' \in C_B[0, \infty) \right\}, \quad \lambda > 0.$$

Theorem 3.18 ([8]) *Let* $f \in H_{n,c}^{(k)} \cap C_b[0, \infty)$. *Then*

$$\|\mathcal{K}_{n,c}^{(k)} f - \mathcal{D}_{n,c}^{(k)} f\| \leq 2K_2 \left(f, \frac{k}{48(n - ck)^2} \right).$$

Along with several interesting results, the authors in [8] obtained quantitative estimates for the difference of Meyer–König Zeller operators (see [124] and [122]) and for Bleimann-Butzer-Hahn operators [30].

Very recently Aral et al. [22] for $f \in C[0, 1]$ considered the Bernstein-type operators of the form:

$$(G_n f)(x) = \sum_{k=0}^{n} e^{-\mu k/n} e^{\mu x} p_{n,k}(a_n(x)) f\left(\frac{k}{n}\right),$$

where $p_{n,k}$ is the Bernstein basis given in (1.1.5), $\mu > 0$ is a real number and

$$a_n(x) = \frac{e^{\mu x/n} - 1}{e^{\mu/n} - 1}.$$

These operators depend on the preservation of exponential functions of the form $\exp_\mu(x) = e^{\mu x}, \mu > 0$. It may be noted that these operators constitute a modification of a special case of the operators due to Morigi and Nematu [131]. These operators are connected with the usual Bernstein polynomials B_n by the following relation:

$$(G_n) f(x) = e^{\mu x} (B_n f_\mu)(a_n(x)), \quad f_\mu = f/\exp_\mu.$$

Pǎltǎnea in [135] (see also [136]) considered the μ-differential and μ-integral , respectively, as $D_\mu : C^1[0, 1] \to C[0, 1]$ and $I_\mu : C[0, 1] \to C^1[0, 1]$ and defined by modified Bernstein–Kantorovich operators as

$$(K^\mu f)(x) = D_\mu \circ B_{n+1} \circ I_\mu,$$

where B_{n+1} is the usual Bernstein polynomials (see (1.1.5)),

$$D_\mu(f, x) = f'(x) - \mu f(x), \ f \in C^1[0, 1], \ x \in [0, 1]$$

and

$$I_\mu(f, x) = e^{\mu x} \int_0^x e^{-\mu t} f(t) dt, \ f \in C[0, 1], \ x \in [0, 1].$$

Aral et al. [24] considered the following modified Kantorovich operators as

$$(\widetilde{K}_n f)(x) = D_\mu \circ G_{n+1} \circ I_\mu.$$

Furthermore, for a given $f \in L_1[0, 1]$, its μ-integral is denoted by $F_\mu \in C[0, 1]$; that is:

$$F_\mu(x) = \int_0^x e^{-\mu t} f(t) dt.$$

The generalized Kantorovich operators (see [24]) can alternatively be defined as

$$(\widetilde{K}_n f)(x) = a'_{n+1}(x)(n + 1) e^{\mu x} \sum_{k=0}^n p_{n,k}(a_{n+1}(x)) \int_{k/(n+1)}^{(k+1)/(n+1)} e^{-\mu t} f(t) dt.$$

Also, the μ-derivative operator satisfies

$$f_\mu^{(r)}(x) = e^{-\mu x} D_\mu^r f(x).$$

Based on the above definitions, Aral-Erbay [22] established the following estimates for the difference of operators:

Theorem 3.19 ([22]) *For $f \in C^r[0, 1]$ and $n > r$, we have*

$$\|D_\mu^r \widetilde{K}_n f - \widetilde{K}_{n-r}(D_\mu^r f)\| \le \frac{\mu}{n+1} e^\mu \sum_{i=0}^{r-1} \|D_\mu^i f\| C_i^r \left(\frac{\mu}{n+1}\right)^{r-i-1} e^{(i+1)\frac{\mu}{n+1}}$$

$$+ \left(\frac{r(r+1)}{2n} e^{\frac{(r+1)\mu}{n+1}} + e^{\frac{(r+1)\mu}{n+1}} - 1\right) e^\mu \|e^{-\mu} D_\mu^r f\|$$

$$+ 2 e^\mu \omega(e^{-\mu} D_\mu^r f, c_n) + 4 e^\mu \omega\left(e^{-\mu} D_\mu^r f, \frac{r}{n-r}\right),$$

where $c_n = (e^{\frac{\mu}{n-r+1}} - e^{\frac{\mu}{n+1}})/(e^{\frac{\mu}{n-r+1}} - 1)$.

Theorem 3.20 ([22]) *For $f \in C^r[0, 1]$ and $n > r$, we have*

$$\|D_\mu^r G_n f - G_{n-r}(D_\mu^r f)\| \leq \frac{\mu}{n} e^\mu \sum_{i=1}^{r-1} \|f_\mu^{(i)}\| \widehat{C}_i^r \left(\frac{\mu}{n}\right)^{r-i-1} e^{i\frac{\mu}{n}}$$

$$+ \left(\frac{r(r-1)}{2n} + e^{\frac{r\mu}{n}} - 1\right) e^\mu \|e^{-\mu} D_\mu^r f\|$$

$$+ 2e^\mu \omega(e^{-\mu} D_\mu^r f, \widehat{c}_n) + e^\mu \omega\left(e^{-\mu} D_\mu^r f, \frac{r}{n-r}\right),$$

where $\widehat{c}_n = (e^{\frac{\mu}{n-r}} - e^{\frac{\mu}{n}})/(e^{\frac{\mu}{n}} - 1)$.

Between the modified Kantorovich variant and the modified Bernstein operators, in ordinary approximation, the following estimate was also provided in [22].

Theorem 3.21 ([22]) *For $f \in C^r[0, 1]$ and $n > r$, we have*

$$\|\widetilde{K}_n f - G_n f\| \leq \left(\frac{\mu}{n+1} - 1\right) e^\mu \|f_\mu\|$$

$$+ e^\mu \left(1 + \left[\left(e^{\frac{2\mu}{n+1}} - 1\right) + 1\right]^n\right) \omega(f_\mu, h_n)$$

$$+ e^\mu \omega(f_\mu, (n+1)^{-1}) + 2e^\mu \omega(f_\mu, v_n),$$

where

$$h_n = \frac{1}{\mu} \left[1 - \frac{2(n+1)}{\mu}\left(e^{\frac{\mu}{n+1}} - 1\right) + \frac{(n+1)}{2\mu}\left(e^{\frac{2\mu}{n+1}} - 1\right)\right]^{1/2}$$

and $v_n = (e^{\frac{\mu}{n-r}} - e^{\frac{\mu}{n+1}})/(e^{\frac{\mu}{n+1}} - 1)$.

Bibliography

1. U. Abel, V. Gupta, Rate of convergence of exponential type operators related to $p(x) = 2x^{3/2}$ for functions of bounded variation. Rev. R. Acad. Cienc. Exactas Fís. Nat. Ser. A Mat. RACSAM **114**, Art 188 (2020)
2. U. Abel, V. Gupta, A complete asymptotic expansion for operators of exponential type with $p(x) = x(1+x)^2$. Positivity **25**, 1013–1025 (2020). https://doi.org/10.1007/s11117-020-00802-5
3. U. Abel, V. Gupta, M. Ivan, On the rate of convergence of Baskakov-Kantorovich-Bézier operators for bounded variation functions. Rev. Anal. Numér. Théor. Approx. **31**(2), 123–133 (2002)
4. U. Abel, V. Gupta, V. Kushnirevych, Asymptotic expansions for certain exponential type operators connected with $2x^{3/2}$. Math. Sci. **15**, 311–315 (2021). https://doi.org/10.1007/s40096-021-00382-9
5. U. Abel, M. Heilmann, V. Kushnirevych, Convergence of linking Baskakov-type operators. Periodica Math. Hungarica **80**, 280–288 (2020)
6. U. Abel, M. Ivan, Some identities for the operator of Bleimann, Butzer and Hahn involving divided differences. Calcolo **36**, 143–160 (1999)
7. U. Abel, M. Ivan, On a generalization of an approximation operator defined by A. Lupaş. Gen. Math. **15**(1), 21–34 (2007)
8. A.M. Acu, S. Hodiş, I. Rasa, Estimates for the differences of certain positive linear operators. Mathematics **8**, 798 (2020). https://doi.org/10.3390/math8050798
9. A.M. Acu, I. Rasa, Estimates for the differences of positive linear operators and their derivatives. Numer. Algorithms **85**, 191–208 (2020)
10. A.M. Acu, I. Rasa, New estimates for the differences of positive linear operators. Numer. Algorithms **73**, 775–789 (2016)
11. A.M. Acu, S. Hodiş, I. Rasa, A survey on estimates for the differences of positive linear operators. Constr. Math. Anal. **1**(2), 113–127 (2018)
12. J.A. Adell, F.G. Badía, J. de la Cal, On the iterates of some Bernstein-type operators. J. Math. Anal. Appl. **209**, 529–541 (1997)
13. O. Agratini, On a sequence of linear and positive operators. Facta Univ. (Niš) Ser. Math. Inform. **14**, 41–48 (1999)
14. O. Agratini, Kantorovich-type operators preserving affine functions. Hacettepe J. Math. Stat. **45**(6), 1657–1663 (2016)
15. P.N. Agrawal, N. Ispir, A. Kajla, Approximation properties of Lupas–Kantorovich operators based on Pólya distribution. Rend. Circ. Mat. Palermo **65**, 185–208 (2016)

16. P.N. Agrawal, A.J. Mohammad, Linear combination of a new sequence of linear positive operators. Revista de la U.M.A. **42**(2), 57–65 (2001)

17. F. Altomare, M. Campiti, *Korovkin-type Approximation Theory and Its Applications*. De Gruyter Studies in Mathematics, vol. 17 (De Gruyter, Berlin, 1994)

18. F. Altomare, S. Diomede, Asymptotic formulae for positive linear operators:direct and converse results. Jaén J. Approx. **2**(2), 255–287 (2010)

19. F. Altomare, I. Raşa, On a class of exponential-type operators and their limit semigroups. J. Approx. Theory **135**, 258–275 (2005)

20. U. Amato, B. Della Vecchia, On Shepard–Gupta-type operators. J. Inequal. Appl. **2018**, 232 (2018). https://doi.org/10.1186/s13660-018-1823-7

21. U. Amato, B. Della Vecchia, Rational approximation on exponential meshes. Symmetry **12**, 1999 (2020). https://doi.org/10.3390/sym12121999

22. A. Aral, H. Erbay, A Note on the difference of positive operators and numerical aspects. Mediterr. J. Math. **17**, 45 (2020). https://doi.org/10.1007/s00009-020-1489-5

23. A. Aral, D. Inoan, I. Rasa, On differences of linear positive operators. Anal. Math. Phys. **9**, 1227–1239 (2019). https://doi.org/10.1007/s1332

24. A. Aral, D. Otroco, I. Raşa, On approximation by some Bernstein–Kantorovich exponential-type polynomials. Periodica Math. Hungarica **79**, 236–254 (2019). https://doi.org/10.1007/s10998-019-00284-3

25. C. Atakut, N. Ispir, Approximation by modified Szász-Mirakjan operators on weighted spaces. Proc. Indian Acad. Sci. Math. **112**, 571–578 (2002)

26. V.A. Baskakov, An instance of a sequence of linear positive operators in the space of continuous functions. Dokl. Akad. Nauk SSSR **113**(2), 249–251 (1957) (In Russian)

27. E.E. Berdysheva, Studying Baskakov Durrmeyer operators and quasi-interpolants via special functions. J. Approx. Theory **149**(2),131–150 (2007)

28. H. Berens, G.G. Lorentz, Inverse theorems for Bernstein polynomials. Indiana Univ. Math. J. **21**, 693–708 (1972)

29. S. Bernstein, Démonstration du théroéme de Weierstrass, fondeé sur le calcul des probabilités. Commun. Soc. Math. Kharkow **13**, 1–2 (1912–1913)

30. G. Bleimann, P.L. Butzer, L. Hahn, A Bernstein-type operator approximating continuous functions of the semi-axis. Indag. Math. **42**, 255–262 (1980)

31. J. Bustamante, Szász-Mirakjan-Kantorovich operators reproducing affine functions. Results Math. **75**, 130 (2020)

32. J. Bustamante, Baskakov-Kantorovich operators reproducing affine functions. Stud. Univ. Babeş-Bolyai Math., to appear

33. W. Chen, On the modified Durrmeyer-Bernstein operator (handwritten, in chinese, 3 pages), in *Report of the Fifth Chinese Conference on Approximation Theory*, Zhen Zhou (1987)

34. C.S. Cismaşiu, About an infinitely divisible distribution, in *Proof of the Colloquium on Approximation and Optimization*, Cluj-Napoca, Oct 25–27 (1984), pp. 53–58

35. C.S. Cismaşiu, On a linear positive operator and its approximation properties. Filomat (Nis) **10**, 159–162 (1996)

36. N.J. Daras, M.Th. Rassias, *Computation, Cryptography, and Network Security* (Springer, Berlin, 2015)

37. N.J. Daras, Th.M. Rassias, *Computational Mathematics and Variational Analysis* (Springer, Berlin, 2020)

38. E. Deniz, A. Aral, V. Gupta, Note on Szász–Mirakyan–Durrmeyer operators preserving e^{2ax}, $a > 0$. Numer. Funct. Anal. Optim. **39**(2), 201–207 (2017)

39. N. Deo, M. Dhamija, D. Miclăuş, Stancu–Kantorovich operators based on inverse Pólya–Eggenberger distribution. Appl. Math. Comput. **273**, 281–289 (2016)

40. N. Deo, S. Kumar, Durrmeyer variant of Apostol-Genocchi-Baskakov operators. Quaestiones Mathematicae (2020). https://doi.org/10.2989/16073606.2020.1834000

41. M.M. Derriennic, Sur lápproximation de fonctions intégrable sur [0, 1] par des polynomes de Bernstein modifiés. Approx. Theory **31**, 325–343 (1981)

42. Z. Ditzian, On global inverse theorems of Szász and Baskakov operators. Canad. J. Math. **31**(2), 255–263 (1979)
43. R.A. DeVore, G.G. Lorentz, *Constructive Approximation* (Springer, Berlin, 1993)
44. M. Dhamija, R. Pratap, N. Deo, Approximation by Kantorovich form of modified Szász–Mirakyan operators. Appl. Math. Comput. **317**, 109–120 (2018)
45. Z. Ditzian, K.G. Ivanov, Bernstein-type operators and their derivatives. J. Approx. Theory **56**, 72–90 (1989)
46. Z. Ditzian, K.G. Ivanov, Strong converse inequalities. J. Anal. Math. **61**, 61–111 (1993)
47. O. Duman, M.A. Ozarslan, B. Della Vecchia, Modified Szász-Mirakjan Kantorovich operators preserving linear functions. Turk. J. Math. **33**, 151–158 (2009)
48. J.L. Durrmeyer, Une formule d' inversion de la Transformee de Laplace, Applications a la Theorie des Moments, These de 3e Cycle, Faculte des Sciences de l' Universite de Paris, 1967
49. Z. Finta, On converse approximation theorems. J. Math. Anal. Appl. **312**, 159–180 (2005)
50. Z. Finta, V. Gupta, Direct and inverse estimates for Phillips type operators. J. Math. Anal. Appl. **303**(2), 627–642 (2005)
51. A.D. Gadjiev, Theorems of the type of P.P. Korovkin's theorems. Math. Zametki **20**(5), 781–786 (1976) (in Russian). Math. Notes **20** (5–6), 995–998 (1976) (Engl. Trans.)
52. S.G. Gal, *Approximation by Complex Bernstein and Convolution type Operators* (World Scientific, Singapore, 2009). https://doi.org/10.1142/7426
53. S.G. Gal, V. Gupta, Quantitative estimates for a new complex Durrmeyer operator in compact disks. Appl. Math. Comput. **218**(6), 2944–2951 (2011). https://doi.org/10.1016/j.amc.2011.08.044
54. S.G. Gal, V. Gupta, Approximation by an exponential-type complex operators. Kragujevac J. Math. **47**(5), 691–700 (2023)
55. S.G. Gal, V. Gupta, N.I. Mahmudov, Approximation by a Durrmeyer-type operator in compact disks. Annali Dell'Univ. Ferrara **58**(2), 65–87 (2012). https://doi.org/10.1007/s11565-011-0124-6
56. W. Gautschi, G. Mastroianni, Th.M. Rassias, *Approximation and Computation, In Honor of Gradimir V. Milovanović* (Springer, Berlin, 2011)
57. H. Gonska, R. Păltănea, Simultaneous approximation by a class of Bernstein-Durrmeyer operators preserving linear functions. Czechoslovak Math. J. **60**(3), 783–799 (2010)
58. H. Gonska, P. Piţul, I. Raşa, On Peano's form of the Taylor remainder, Voronovskaja's theorem and the commutator of positive linear operators, in *Numerical Analysis and Approximation Theory (Proceedings of the 2006 International Conference held at Cluj-Napoca)*
59. H. Gonska, M. Heilmann, I. Rasa, Kantorovich operators of order k. Numer. Funct. Anal. Optimiz. **32**(7), 717–738 (2011)
60. T.N.T. Goodman, A. Sharma, A modified Bernstein-Schoenberg operator, in ed. by Bl. Sendov, *Constructive Theory of Functions*, Varna, 1987 (Public House Bulgarian Academy of Sciences, Sofia, 1988), pp. 166–173
61. N.K. Govil, V. Gupta, D. Soybaş, Certain new classes of Durrmeyer type operators. Appl. Math. Comput. **225**, 195–203 (2013)
62. V. Gupta, Higher order Lupaş-Kantorovich operators and finite differences. Rev. R. Acad. Cienc. Exactas Fís. Nat. Ser. A Mat., RACSAM **115**, 100 (2021). https://doi.org/10.1007/s13398-021-01034-2
63. V. Gupta, Convergence of exponential operators connected with x^3 on functions of BV, Miskolc Math. Notes, to appear
64. V. Gupta, A general class of integral operators. Carpathian J. Math. **36**(3), 423–431 (2020)
65. V. Gupta, Convergence estimates for Gamma operator. Bull. Malaysian Math Soc. **43**(3), 2065–2075 (2020)
66. V. Gupta, Approximation with certain exponential operators. Rev. R. Acad. Cienc. Exactas Fís. Nat. Ser. A Mat. RACSAM **114**(2), 51 (2020)

67. V. Gupta, Convergence estimates of certain exponential type operators, in ed. by Deo et al., *Mathematical Analysis I – Approximation Theory*. Springer Proceedings in Mathematics & Statistics, vol. 306 (Springer, Berlin, 2020), pp. 47–56

68. V. Gupta, A large family of linear positive operators. Rend. Circ. Mat. Palermo, II. Ser **69**, 701–709 (2020). https://doi.org/10.1007/s12215-019-00430-3

69. V. Gupta, Estimate for the difference of operators having different basis functions. Rend. del Circ. Mat. di Palermo Ser. **69**, 995–1003 (2020). https://doi.org/10.1007/s12215-019-00451-y

70. V. Gupta, On difference of operators with applications to Szász type operators. Rev. R. Acad. Cienc. Exactas Fís. Nat. Ser. A Mat., RACSAM **113**, 2059–2071 (2019)

71. V. Gupta, Differences of operators of Baskakov type. Fasciculi Math. **62**, 47–55 (2019)

72. V. Gupta, A note on general family of operators preserving linear functions. Rev. R. Acad. Cienc. Exactas Fís. Nat. Ser. A Mat., RACSAM **113**(4), 3717–3725 (2019)

73. V. Gupta, Some examples of genuine approximation operators. General Math. **26**(1–2), 3–9 (2018)

74. V. Gupta, Differences of operators of Lupaş type. Construct. Math. Analy. **1**(1), 9–14 (2018)

75. V. Gupta, Approximation properties by Bernstein-Durrmeyer type operators. Complex Anal. Oper. Theory **7**, 363–374 (2013)

76. V. Gupta, A.M. Acu, On difference of operators with different basis functions. Filomat **33**(10), 3023–3034 (2019)

77. V. Gupta, A.M. Acu, H.M. Srivastava, Difference of some positive linear approximation operators for higher-order derivatives. Symmetry **12**(6), Art. 915 (2020)

78. V. Gupta, R.P. Agarwal, *Convergence Estimates in Approximation Theory* (Springer, Berlin, 2014)

79. V. Gupta, G. Agrawal, Approximation for link Ismail-May operators. Ann. Funct. Anal. **11**, 728–747 (2020). https://doi.org/10.1007/s43034-019-00051-y

80. V. Gupta, D. Agarwal, M.Th. Rassias, Quantitative estimates for differences of Baskakov-type operators. Complex Anal. Oper. Theory **13**(8), 4045–4064 (2019)

81. V. Gupta, J. Bustamante, Kantorovich variant of Jain-Pethe operators. Numer. Funct. Anal. Optimiz. **42**, 551–566 (2021). https://doi.org/10.1080/01630563.2021.1895834

82. V. Gupta, O. Duman, Bernstein Durrmeyer type operators preserving linear functions. Math. Vesniki **62**(4), 259–264 (2010)

83. V. Gupta, M.K. Gupta, V. Vasishtha, Simultaneous approximation by summation integral type operators. J. Nonlinear Funct. Analy. Appl. **8**(3), 399–412 (2003)

84. V. Gupta, G.C. Greubel, Moment estimations of a new Szász-Mirakyan-Durrmeyer operators. Appl. Math Comput. **271**, 540–547 (2015)

85. V. Gupta, P. Maheshwari, Bézier variant of a new Durrmeyer type operators. Rivista di Matematica della "Università di Parma" **7**(2), 9–21 (2003)

86. V. Gupta, M.A. Noor, Convergence of derivatives for certain mixed Szász–Beta operators. J. Math. Anal. Appl. **321**(1), 1–9 (2006)

87. V. Gupta, M. López-Pellicer, H.M. Srivastava, Convergence estimates of a family of approximation operators of exponential type. Filomat **34**(13), 4329–4341 (2020)

88. V. Gupta, C.P. Muraru, V.A. Radu, Convergence of certain hybrid operators. Rocky Mountain J. Math. **51**(4), 1249–1258 (2021)

89. V. Gupta, M.Th. Rassias, *Moments of Linear Positive Operators and Approximation*. SpringerBriefs in Mathematics (Springer, Cham, 2019). https://doi.org/10.1007/978-3-030-19455-0

90. V. Gupta, M.Th. Rassias, Asymptotic formula in simultaneous approximation for certain Ismail-May-Baskakov operators. J. Numer. Anal. Approx. Theory, to appear.

91. V. Gupta, Th.M. Rassias, P.N. Agrawal, A.M. Acu, Estimates for the differences of positive linear operators, in *Recent Advances in Constructive Approximation Theory*. Springer Optimization and Its Applications, vol. 138, (Springer, Cham, 2018)

92. V. Gupta, Th.M. Rassias, Lupaş-Durrmeyer operators based on Pólya distribution. Banach J. Math. Anal. **8**(2),146–155 (2014)

93. V. Gupta, G.S. Srivastava, Simultaneous approximation by Baskakov-Szász type operators. Bull. Math.de la Soc. Sci. de Roumanie (N. S.) **37**(85), No. 3–4, 73–85 (1993)

94. V. Gupta, R. Yadav, On approximation of certain integral operators. Acta Math. Vietnam **39**, 193–203 (2014)

95. V. Gupta, G. Tachev, Some results on Post-Widder operators preserving test function x^r. Kragujevac J. Math. **46**(1), 149–165 (2022)

96. V. Gupta, G. Tachev, A modified Post Widder operators preserving e^{Ax}. Stud. Univ. Babeş-Bolyai Math., in press.

97. V. Gupta, G. Tachev, *Approximation with Positive Linear Operators and Linear Combinations*. Developments in Mathematics, vol. 50 (Springer, Berlin, 2017)

98. V. Gupta, G. Tachev, A note on the differences of two positive linear operators. Construct. Math. Analy. **2**(1), 1–7 (2019)

99. M. Herzog, Semi-exponential operators. Symmetry **13**, 637 (2021). https://doi.org/10.3390/sym1304063

100. M. Heilmann, M.W. Müller, On simultaneous approximation by the method of Baskakov-Durrmeyer operators. Numer. Funct. Anal. Optimiz. **10**(1–2), 127–138 (1989)

101. M. Heilmann, I. Raşa, A nice representation for a link between Baskakov and Szász–Mirakjan–Durrmeyer operators and their Kantorovich variants. Results Math. **74**, 9 (2019). https://doi.org/10.1007/s00025-018-0932-4

102. A. Holhoş, The rate of approximation of functions in an infinite interval by positive linear operators. Stud. Univ. Babeş-Bolyai Math. **2**, 133–142 (2010)

103. M. Ismail, C.P. May, On a family of approximation operators. J. Math. Anal. Appl. **63**, 446–462 (1978)

104. N. Ispir, On modified Baskakov operators on weighted spaces. Turk. J. Math. **26**(3), 355–365 (2001)

105. N. Ispir, I. Yuksel, On the Bézier variant of Srivastava-Gupta operators. Appl. Math E-Notes **5**, 129–137 (2005)

106. M. Ivan, A note on the Bleimann-Butzer-Hahn operator. Automat. Comput. Appl. Math. **6**, 11–15 (1997)

107. G.C. Jain, Approximation of functions by a new class of linear operators. J. Austral. Math. Soc. **13**, 271–276 (1972)

108. G.C. Jain, S. Pethe, On the generalizations of Bernstein and Szász-Mirakyan operators. Nanta Math. **10**, 185–193 (1977)

109. A. Kajla, Direct estimates of certain Miheşan-Durrmeyer type operators. Adv. Oper. Theory **2**(2), 162–178 (2017)

110. L.V. Kantorovich, Sur certains developements suivant les polynômes de la forme de S. Bernstein I, II. Dokl. Akad. Nauk SSSR **563**(568), 595–600 (1930)

111. Km. Lipi, N. Deo, General family of exponential operators. Filomat **34**(12), 4043–4060 (2020)

112. Km. Lipi, N. Deo, On modification of certain exponential type operators preserving constant and e^{-x}. Bull. Malays. Math. Sci. Soc. (2021). https://doi.org/10.1007/s40840-021-01100-3

113. A.-J. López-Moreno, Weighted simultaneous approximation with Baskakov type operators. Acta Math. Hungar. **104**(1–2), 143–151 (2004)

114. A. Lupaş, The approximation by means of some linear positive operators, in ed. by M.W. Muller et al., *Approximation Theory* (Akademie-Verlag, Berlin, 1995), pp. 201–227

115. A. Lupaş, Die Folge Der Betaoperatoren, Dissertation, Universitat Stuttgart, 1972

116. L. Lupaş, A. Lupaş, Polynomials of binomial type and approximation operators. Studia Univ. Babes-Bolyai Math. **32**, 61–69 (1987)

117. L. Lupaş, A. Lupaş, Polynomials of binomial type and approximation operators. Studia Univ. Babeş-Bolyai Math. **32**(4), 61–69 (1987)

118. C.P. May, Saturation and inverse theorems for combinations of a class of exponential-type operators. Canad. J. Math. **28**, 1224–1250 (1976)

119. C.P. May, On Phillips operator. J. Approx. Theory **20**(4), 315–332 (1977)

120. G. Mastroianni, Su una classe di operatori lineari e positivi. Rend. Acc. Sc. Fis. Mat., Napoli **48**(4), 217–235 (1980)

121. V. Miheşan, Gamma approximating operators. Creative Math. Inf. **17**, 466–472 (2008)

122. V. Maier, M.W. Müller, J. Swetits, The local L_1 saturation class of the method of integrated Meyer-König and Zeller operators. J. Appr. Theory **32**, 27–31 (1981)

123. S.M. Mazhar, V. Totik, Approximation by modified Szász operators. Acta Sci. Math. (Szeged) **49**, 257–269 (1985)

124. W. Meyer-König, K. Zeller, Bernsteinsche Potenzreihen. Studia Math. **19**, 89–94 (1960)

125. D. Miclăuş, The revision of some results for Bernstein-Stancu type operators. Carpathian J. Math. **28**(2), 289–300 (2012)

126. G.V. Milovanović, M.Th. Rassias, *Analytic Number Theory, Approximation Theory and Special Functions* (Springer, Berlin, 2014)

127. G.M. Mirakjan, Approximation of continuous functions with the aid of polynomials (Russian). Dokl. Akad. Nauk SSSR **31**, 201–205 (1941)

128. N.S. Mishra, N. Deo, Kantorovich variant of Ismail–May operators. Iran J. Sci. Technol. Trans. Sci. **44**, 739–748 (2020). https://doi.org/10.1007/s40995-020-00863-x

129. N.S. Mishra, N. Deo, On the preservation of functions with exponential growth by modified Ismail–May operators. Math. Methods Appl. Sci. (2021). https://doi.org/10.1002/mma.7328

130. A.J. Mohammad, A.K. Hassan, Simultaneous approximation by a new sequence of Szász-Beta type operators. Rev. de la un. Mat. Argentina **50**(1), 31–40 (2009)

131. S. Morigi, M. Neamtu, Some results for a class of generalized polynomials. Adv. Comput. Math. **12**(2–3), 133–149 (2000)

132. M.W. Müller, L_p approximation by the method of integral Meyer König and Zeller operators. Studia Math. **63**, 81–88 (1978)

133. J. Nagel, Kantorovich operators of second order. Monatsh. Math. **95**, 33–44 (1983)

134. T. Neer, P.N.Agrawal, A genuine family of Bernstein-Durrmeyer type operators based on Pólya basis functions. Filomat **31**(9), 2611–2623 (2017)

135. R. Păltănea, A note on generalized Benstein-Kantorovich operators. Bull. Transilvania Univ. Brasov, Ser. III Math. Infor. **6**(55)(2), 27–32 (2013)

136. R. Păltănea, A generalization of Kantorovich operators and a shape-preserving property of Bernstein operators. Bull. Transilvania Univ. Brasov Ser. III Math. Infor. Phys. **5**(54)(2), 65–68 (2012)

137. R. Păltănea, Estimates of approximation in terms of a weighted modulus of continuity. Bull. Transilvania Univ. of Brasov **4**(53), 67–74 (2011)

138. R. Păltănea, Modified Szász-Mirakjan operators of integral form. Carpathian J. Math. **24**(3), 378–385 (2008)

139. R. Păltănea, A class of Durrmeyer type operators preserving linear functions. Ann. Tiberiu Popoviciu Sem. Funct. Equat. Approxim. Convex. (Cluj-Napoca) **5**, 109–118 (2007)

140. E. Pandey, R.K. Mishra, Direct estimates for Gupta type operators. J. Classical Analy. **17**(1), 27–37 (2021)

141. R.S. Phillips, An inversion formula for Laplace transformation and semi- groups of linear operators. Ann. Math. **59**, 325–356 (1954)

142. G. Prasad, P.N. Agrawal, H.S. Kasana, Approximation of functions on $[0, \infty)$ by a new sequence of modified Szász operators. Math. Forum **6**(2), 1–11 (1983)

143. A. Raigorodskii, M.Th. Rassias, *Trigonometric Sums and their Applications* (Springer, Berlin, 2020)

144. M.Th. Rassias, *Harmonic Analysis and Applications* (Springer, Berlin, 2021)

145. R.K.S. Rathore, P.N. Agrawal, Inverse and sturation theorems for derivatives of exponential type operators. Indian J. Pure. Appl. Math. **13**(4), 476–490 (1982)

146. Q. Razi, Approximation of a function by Kantorovich type operators. Mat. Vesnic. **41**, 183–192 (1989)

147. A. Sahai, G. Prasad, On simultaneous approximation by modified Lupas operators. J. Approx. Theory **45**(2), 122–128 (1985)

148. K. Sato, Global approximation theorems for some exponential-type operators. J. Approx. Theory **32**, 32–46 (1981)

149. O. Shisha, B. Mond, The degree of convergence of linear positive operators. Proc. Nat. Acad. Sci. USA **60**, 1196–1200 (1968)

150. D. Soybaş, N. Malik, Convergence estimates for Gupta-Srivastava operators. Kragujevac J. Math. **45**(5), 739–749 (2021)

151. H.M. Srivastava, V. Gupta, A certain family of summation-integral type operators. Math. Comput. Modelling **37**, 1307–1315 (2003)

152. D.D. Stancu, Approximation of functions by a new class of linear polynomial operators. Rev. Roum. Math. Pures et Appl. **13**, 1173–1194 (1968)

153. O. Szász, Generalizations of S. Bernstein's polynomials to the infinite interval. J. Res. Nat. Bureau Standards **45**(3), 239–245 (1950)

154. G. Başcanbaz-Tunca, M. Bodur, D. Söylemez, On Lupaş-Jain Operators. Stud. Univ. Babeş-Bolyai Math. **63**(4), 525–537 (2018)

155. A. Tyliba, E. Wachnicki, On some class of exponential type operators. Comment. Math. **45**, 59–73 (2005)

156. S. Umar, Q. Razi, Approximation of function by generalized Szász operators. Commun. Fac. Sci. de lÚniversité dÁnkara, Serie A1: Math. **34**, 45–52 (1985)

157. D.V. Widder, *The Laplace Transform*, Princeton Mathematical Series (Princeton University Press, Princeton, 1941)

158. X.M. Zeng, U. Abel, M. Ivan, A Kantorovich variant of the Bleimann, Butzer and Hahn operators. Math. Inequal. Appl. **11**(2), 317–325 (2008)

159. C. Zhang, Z. Zhu, Preservation properties of the Baskakov–Kantorovich operators. Comput. Math. Appl. **57**(9), 1450–1455 (2009)

Printed in the United States
by Baker & Taylor Publisher Services